Biopsychology and Your Mind

Biopsychology and Your Mind explains the way your mind works – and why it sometimes doesn't. The key is your mental mechanics: molecular, neural, and biological mechanisms.

This book explores the unexpected properties of thoughts, feelings, emotions, and actions, examining phenomena such as blindsight, associative learning, intact memory in amnesia, and dreaming during sleep. Using evolutionary biopsychology as a framework, it addresses fundamental questions: What constitutes an emotion? How do thoughts and emotions interact? This text reveals how evolution has shaped our mental mechanics to seek adaptive balances, often switching between extremes depending on current circumstances. Mental disorders are presented as otherwise adaptive reactions that occur either in unusually extreme situations or within systems that are unusually reactive. This book demonstrates how evolutionary pressures have created mental systems that rapidly change between states, seeking optimal responses rather than fixed solutions.

This book is a must-read for anyone seeking to understand mental machinery and psychological disorders. Detailed notes, references, and technical sections also make it an ideal companion for all students of biopsychology.

Neil McNaughton is an Emeritus Professor and Fellow of the Royal Society of New Zealand in the Department of Psychology (Te Tari Whakamātau Hinekaro) at the University of Otago (Ōtākou Whakaihu Waka) in Aotearoa/New Zealand.

Biopsychology and Your Mind

Biopsychology and Your Mind

The Mental Mechanics of Your Thoughts, Feelings, Emotions, and Their Disorders

Neil McNaughton

Routledge
Taylor & Francis Group
LONDON AND NEW YORK

Designed cover image: Getty Images

First published 2026
by Routledge
4 Park Square, Milton Park, Abingdon, Oxon OX14 4RN

and by Routledge
605 Third Avenue, New York, NY 10158

Routledge is an imprint of the Taylor & Francis Group, an informa business

British Library Cataloguing-in-Publication Data
A catalogue record for this book is available from the British Library

ISBN: 9781032735986 (hbk)
ISBN: 9781032735962 (pbk)
ISBN: 9781003465027 (ebk)

DOI: 10.4324/9781003465027

Typeset in Sabon
by codeMantra

Contents

Preface

My daughter Catriona was born, and my book *Biology and Emotion* was sent to the publisher, in 1988. Eight years later:

> "Dad, write a book for me."
> "What sort of book?"
> "Write a book that explains emotions to me."

At age 8, Catriona had no idea of how emotions worked, nor how slowly her dad wrote books. During the next 30 years, I published the second and third editions of *Neuropsychology of Anxiety*. But, somehow, I had still not completed her book. It turns out that she then wrote (as a writing exercise):

> I have a clear memory of picking up a copy of Biology of Emotion, seeing the acknowledgement to my mum and brother and questioning my absence. I remember sitting on my dad's knee and asking him to explain both my absence and its contents. I can also clearly picture standing in our attic, seeing an illusion puzzle in a book and connecting this to the centre of what MY book would be about. I'm also sure that these aren't 'real' memories. I am sure I have created them.
>
> Knowing that I don't know, and that 'my' perception is just 'a' perception, has shaped my life.
>
> I am fascinated both with how much we know and don't know about our brains. To me, it is like the ocean, which I grew up looking at every day. So familiar, but the depths so unknown.
>
> We have picked up [the writing of] this book over the years, and I know I have recreated those memories because of it. The *Neuropsychology of Anxiety* (third edition), which finally included my acknowledgement, was published in 2024.
>
> ... This new book opens a window.

This book is for you too, if, like Catriona, you are "fascinated both with how much we know and don't know about our brains" and with how little we truly understand of its workings. So, I feel it's past time to explain in simple everyday terms, to Catriona and everyone else, what I currently know of how the mind works after over 50 years of study.

Why emphasise emotion? Emotional reactions inextricably involve perceptions, thoughts, feelings, desires, and actions – most of what the brain does. To explain what I understand of emotion, I found I had to explain all of that too. What I have at last delivered was meant to be 'a book that explains emotions' for Catriona and, later, you. But, of necessity, it now covers the biopsychology of the mind as a whole. A final twist was my publisher wanting it to be usable by first-year university students as well – hence references and notes that you can happily ignore.

Critically, it is emotion that gives you a basis for action. Without emotion, there would be no reason to act or need to make sense of your actions. To totally control your emotions, to limit them, and constrain them, might make your life less painful, but, like any anaesthetic, it would make it a very poor and dull thing. Conversely, insufficient control is where disorder and disaster lie. Mental mechanics should help you understand the necessary balances (which underlie the evolutionary necessity for emotion in the first place) and begin to give you the tools to achieve them.

Whether you are 'neurotypical' or not, this book should help you understand who and how you are. At the brain level, we are all different, but the rules of our mental processes are the same. Your mind or that of someone close to you can cause problems. Understanding the processes can then become urgent. So, we end with a chapter on, so-called, mental disorders. I am not a fan of the term 'disorder', here. It implies some specific pathology. In contrast, mental problems are usually just extreme or unwanted versions of healthy adaptive reactions. Usually, you will seek help because of the emotions directly or indirectly involved, but with schizophrenia, for example, the problems also arise from unwanted cognitions – which become problematic because of the consequent emotions.

I hope that after reading this book, you will see the tapestry of your thoughts and feelings as richer, and more colourful, than before (Chapter 1). I hope you will see that even your most disturbing emotions have potential value (Chapter 10). They are the result of a huge store of evolutionary experience that is often wiser than your cold 'rational' thought (Chapters 2–9); although they can get out of hand.

Paths through this book

So, here is a picture of what I have come to call Mental Mechanics. It's mainly painted for *anyone* interested in the workings of their own mind – and particularly their emotions.

I have flagged some more technical sections (and whole chapters) that you can skip if you want to avoid technical detail without losing the plot. Importantly, you can easily go back to skipped sections and chapters (particularly 3 and 4, which are on how the brain is built) when you feel a need. It also has endnotes and references. These are cited in superscript (references have an 'R' in front) and listed at the end of each chapter.[1] Most of the endnotes are technical expansions or follow-on comments, and you can ignore them unless you are particularly interested or on your second read through. If you are a first-year undergraduate, they should give you stepping-stones to a higher-level biopsychology course.

You can read straight through in the given chapter order. They tell a coherent story. But you can also take a number of different paths through the chapters, following your fancy. Most chapters can be read stand-alone. So, pick from them what interests you most and skip forward and backward as you like and as you feel the need. I have tried to make a story out of the science – but science itself is more like a spider's web. You can always go back to an earlier chapter or section. A number of electronic links are included that offer you similar choices with material that is outside this book. With the electronic version of the book, you can jump to and from these links and to and from the endnotes.

Neil McNaughton

Note

1 This is an example of an endnote. This one provides a note on abstracts for those not used to seeing them. Each chapter in the electronic version of the book will start with an abstract, a short summary of its content. The abstract should help you to decide if you want to explore this chapter now or skip it and come back later. The abstracts, endnotes, and references have all been added for undergraduate students as a springboard to higher-level material. If you want to go to that level, follow them. If you just want the big picture, don't.

Acknowledgements

As described in the preface, this book was specifically requested by Catriona McNaughton three decades ago. She helped draft initial versions of the first two chapters and provided detailed corrections and advice on the final versions of Chapters 1 and 2. I am particularly grateful for her patience with me over this long period.

The final text summarises material that I refined over more than 30 years of teaching *Biopsychology* at the University of Otago. The teaching involved was very much two-way. Every year my students would ask questions, make comments and specific suggestions, and even offer me without prompting their own calculations of things like the brain's computational capacity in digital terms. (An early answer was a 3-metre stack of DVDs, updated by a later student to a 20 cm stack of 1 TB hard drives.) In a number of cases, they provided critiques, and alternative suggestions, that caused me to change the textbook used in the following years. My thanks go to all of them for shaping my thinking and, in particular, encouraging me to use simple in-class experiments to help their understanding.

I am also grateful for the specialist input provided over the years by other members of my department and by members of the medical school on topics such as long-term potentiation and reinforcement. I would particularly like to take this opportunity to thank our wonderful departmental technicians (especially Lindsay Robertson) not only for their superb help in supporting my research but also for their willingness to develop and build specialised rigs for my in-class demonstrations based on their close engagement with the students. The writing of this book was supported by the continued provision of office and library resources and technical support by the University of Otago.

Twenty years ago, Fred Toates advised me strongly not to try and write a conventional biopsychology textbook given his experience of doing so. This influenced both what sort of book this is and how long it was before it was written. Robin Dunbar interested me in psychology when I found my chemistry degree boring; and, later, convinced me of the value of evolutionary adaptation in psychological arguments. I have been strongly influenced by his idea of the evolution of language as an extension of grooming that allows large social group size. It may not be obvious, but this has set the tone for many of the arguments in this book.

The writing of this book has benefited from incisive comments on earlier drafts by Alison McNaughton (Chapter 2), Owen Jones (Chapter 2), and Cliff Abraham (Chapter 3). Drew Maurer read Chapters 3 and 4. For Chapter 10, John Williams provided psychiatric comment and Matt Stevenson and Damon Mitchell provided clinical psychology comment and stylistic changes. Matt also kindly gave me permission to use his own experience as a case example. Last but not least, my long-time collaborator Philip Corr *spontaneously* volunteered to read the entire manuscript. It has benefited throughout from his detailed detection of textual errors and his suggestions as to content, strategy, and phrasing. I have done my best to follow the advice I have been given by everyone, but this has not always been within my abilities, and so the remaining deficiencies are definitely mine.

Finally, I would like to thank Ceri McLardy and her team at Routledge. They solved a number of problems in organising the text that arose over the years – always in the direction of simpler implementation than I had suggested. Ceri was instrumental in choosing the cover.

1 The multi-hued mind

Why did I do that?

If you are like me, your own mind can surprise you. You may do something unexpected. Often you drive on your work route, when you were going somewhere else. As you left home, you started thinking about something else.

Later you may wonder: Where does my mind come from? How does my mind work? How does my brain do this? Your mind cannot be your brain. Nor is the mind a program that runs in the brain's computational circuits. The mind emerges from brain activity. Do not confuse the hardware (actually wetware) with the software or its *output*. But do not downplay the mechanics of the wetware in generating the (experiential and real-world) output.

Biopsychology and Your Mind explains the fundamentals of *how* your mind works in terms of the mechanisms of your brain. Basic 'mental mechanics' should help you answer the fancy questions like "What is mind?".[1] They should also help you answer the smaller bread-and-butter questions, which are much more important on a day-to-day basis, like "Why do I feel so rotten and how can I change it?".

"Why did they...?" and "Will they...?" can seem even more important questions than "Why did I...?". Your job, or being loved, or just having a good time, can depend on understanding someone else's mind. I will not give you specific answers to these questions, especially as I don't know whose mind you particularly want to understand. But, in science, if you understand general principles, you can solve your particular problems. This book is heavy on principles – details can be chased up later (see Chapter 11).

Biopsychology and Your Mind explains the invisible, often counter-intuitive, basic processes underlying emotion and thinking (including intuition). What it tells you applies to both you and others; to both well and unwell or damaged minds. Your mind normally works so smoothly that only when another mind is broken do you get a hint of what your own mind is doing.

Understanding how a mind gets broken may help you understand a broken mind. But this echoes into your own thoughts and, particularly, emotions. Starting with the next chapter, throughout this book, I will try to provide you

DOI: 10.4324/9781003465027-1

with demonstrations that temporarily distort your mind to show you how it works; and with illusions that show how it sometimes doesn't work. Even a well mind is imperfect: distorting 'reality' all the time. Paradoxically, this often improves survival.

Emotions are one of the main sources of 'Why…?' questions about ourselves and others. 'Mental mechanics' can sound cold, dry, computational. But we will ask about the great emotions, such as love – and its difference from lust. Our minds evolved to achieve goals. 'Cognitively processing' without emotion would not lead to action. Of course, we also need to look at how we see the world and navigate problems. But, if pure thought exists, it is not what our minds are *for*.

Emotions are like weather in your head. Positive feelings are like internal rainbows. To enjoy a real rainbow is not to ask 'why' or 'how'. So, should you ask *why* the rainbow in your head exists? Scientific analysis could destroy its beauty.[R2] But, before you stop reading, just think about bad weather, too.

Negative feelings are like bad mental weather. Not only tumultuous but shutting out the light. When you ask "*Why* do I feel like this?", it's likely to be due to confusion or emotional agony.[2] Understanding what generates such feelings – understanding their mental mechanics – *why* they occur at all – will help you deal with them. Contrast dark and light. It will help you see them as distinct, functional, and part of your world. But they both derive from the same mental mechanics. Understand this and you will come to see your mind (in gloom or joy) as multi-hued, rich, and wonderful. There are depths in you that you cannot experience but can understand through the lens of neuroscience.

This book tackles a follow-on question. How can simple physical processes lead to such a complex and compelling psychological experience? How do people see the colours they see? What, fundamentally, is colour as an experience rather than as a physical parameter? Why does seeing colour, particularly in certain combinations, produce emotions that cause people to spend vast sums of money on particular arrangements of hues? At this point, scientific knowledge can impact art and help us, as artists, manipulate others' experience – particularly in approaches such as op art (Figure 1.1).

Our *experience* of an externally generated rainbow is, of course, inside us but only one step from the physics of diffraction. But our thoughts and emotions, unlike raindrops and sunlight, are not easy to detect or change systematically. So, the mind is hard to study. But it is understandable in terms of high-level mental mechanical principles even without precise low-level details.

Our experience of emotions is particularly mysterious. They are the result of changes (including bodily changes) inside you. So, both cause and effect are internal and harder to study than external events. But even emotion can be understood with mental mechanical principles that account for our *experience* of the rainbow.

Figure 1.1 Seeing what is not there. Lines close together viewed from the right distance tend to shimmer, especially if they are curved rather than straight. This is due to the way our brain processes visual information (see Chapter 9).

'Mental mechanics' in my subtitle deliberately echoes 'quantum mechanics'. Both involve quite simple basic principles. But, if you apply them, they lead to major departures *from our everyday views of how the world works.* There is also a sense in which I will be describing the fundamental 'particles' of mental processes. We will look at the way that free will may reflect a quantum mechanical-like uncertainty within human decision making, and note that at least one theory treats consciousness as deriving from quantum physics. But, despite dealing with such principles, the important result is that mental mechanics highlights and goes some way to explain, all the magnificent weirdness we see. It underpins the departures from rationality that make our friends fascinating and that have strongly contributed to recent economic crises. The predictive failures of classical "human choices are rational" economics have forced the development of neuroeconomics.[R3–6] Unfortunately, politicians still use classical economics.

This book is meant to give you an idea of how to take your mind apart, and how to understand the parts. It aims to tell you how your mind – the sum total of your conscious and unconscious thoughts, feelings, and emotions – works; and why your mind sometimes fails to work the way you would wish. *Biopsychology and Your Mind* provides a starting point for an exploration of healthy psychological mechanisms and a basis for understanding mental disorders.

A key feature of *Biopsychology and Your Mind* is that it treats the mind as a product of brain activity. This allows us to know, for example, that your dreams involve your mind being awake even when your body looks asleep

(Chapter 8). Likewise, mind blanking involves your mind being asleep when your body looks awake.[R1] The next chapter explains why you should always see your brain as the key to all your mental faculties.

Experiment 1.1 [Part A]

Before you move on, as the first step in an experiment we will complete and explain later, please look at the photographs below of three out of four 'quadruplets'. (We will meet the fourth one later.)

Neil

Alastair

Ian

Write down on a slip of paper whether you think Neil looks more like Alastair or more like Ian. Push the slip with your answer in where the middle of this page meets the binding so it will be held tightly. We'll come back for it later. Don't be tempted to actually write on the book – you may want to check out some friends' responses later and see if they are the same or different. I'll explain later how you can run a full version of this experiment, as a party trick, with a group of friends.

Notes

1 A question perhaps currently best answered (Chapter 9) with the joke: "What is mind? No matter. What is matter? Never mind". Title of a review by Nicholas Humphrey in the *London Review of Books*, Vol. 3, No. (24–17 December 1981) of *The Mind's I: Fantasies and Reflections on Self and Soul*, ed. Douglas Hofstadter and Daniel Dennett.

2 We will see later that emotional pain activates the same areas of the brain that register the pain caused by physical damage, and to the same extent. So, we are

justified in talking about emotional agony and seeing it as just as real and problematic and as in need of help as physical agony.

References

R1. Andrillon, T, Lutz, A, Windt, J, & Demertzi, A, Where is my mind? A neurocognitive investigation of mind blanking. *Trends in Cognitive Sciences*, 2025. 29VFV: 600–13.
R2. Dawkins, R, *Unweaving the Rainbow: science, delusion and the appetite for wonder*. 2000: Allen Lane, The Penguin Press, London.
R3. Kenning, P & Plassmann, H, NeuroEconomics: An overview from an economic perspective. *Brain Research Bulletin*, 2005. **67**: 343–54.
R4. Loewenstein, G, Scott, R, & Cohen, JD, Neuroeconomics. *Annual Review of Psychology*, 2008. **59**: 647–72.
R5. Stuphorn, V, Neuroeconomics: The shadow of the future. *Current Biology*, 2005. **15**: R247–9.
R6. Zak, PJ, Neuroeconomics. *Philosophical Transactions of the Royal Society of London: B*, 2004. **359**: 1737–48.

2 The mind machine

From mind to brain

Your mind is special, personally and scientifically. It's the only one that you can be absolutely sure exists. Of course, if asked, I would say that I have a mind too; but I would say that, wouldn't I? Even a robot with a processor built of cogs and wheels (e.g., The Babbage Engine https://www.computerhistory.org/babbage/) would say it had a mind if it was designed to do so.[1]

This creates a problem for scientists. How can we tell you how your mind works if we can't be certain that you even have one? The answer is to look at parts of the problem, then put them together to make a complete picture. Let me introduce you to the mind machine – your brain. I know your mind and brain have shared the same living quarters all your lives, but I think they may not have been properly introduced to each other.

Focus on your brain. You don't need to decide if anyone, other than you, has a mind, or what a mind is. If I can convince you that all of your mind depends on your brain, showing you your mental mechanics, we can ultimately work out what your mind really is. You may be surprised by what you find in there as we go along.

Critically, we can take a brain out and look at it. We can see that everyone's brain has similar machinery[2]; and that this machinery shares features with the brains of other animals, too. The details vary across your lifespan, between people, and between species. Your reactions to specific stimuli can vary from moment to moment. But the basic ground rules – the mental mechanics – of brain architecture and processing are the same for everyone from mollusc[3] to mouse to me. I'll be trusting in the similarity of your brain and mine when giving you experiments to test how your own mind works.

Going brain step by brain step into the mind, we will see how the mind changes when we change the brain. We will find that apparently seamless mental processes, like seeing, are actually patchworks. They involve multiple, parallel, hierarchically organized processes.

DOI: 10.4324/9781003465027-2

Perception

I'll use vision to show you how perceptual systems work. Different sensory inputs produce different, independent reactions.[4] But their mental mechanics are fundamentally the same.

To start, let's see how the brain game is played with one of the simplest possible experiments with your visual system. I'll put the experiments into boxes separate from the text, but this one, Experiment 2.1, is so simple that it only needs one step.

Experiment 2.1 A temporary visual system lesion

Close your eyes or cover them with a high-quality sleep mask or completely darken the room you are in. The key experimental result is that you no longer 'see' your surroundings. (Notice that I am pretty sure of this – but in the science game, it's up to you to check that I am right when I tell you what you should observe.) You are also likely to 'see' a range of often coloured and often geometric moving patterns, but they are completely abstract and don't look like objects or a normal view of the world. I see these patterns as inside my head rather than in my surroundings – but you may not. The key thing is that you can't see your surroundings, and so the patterns must be brain-generated. Even with input from your surroundings, your perceptions are generated by brain processing.

This simple experimental manipulation (which you can think of as a temporary lesion of your visual input) shows that for you to see the world light must enter your eyes; and your eyes must then send information to your brain.

In experiments like this it is a good idea to include positive as well as negative controls. Your visual contact with the world has gone, but (again I am sure) you know the world is still there because you can still hear, feel, and navigate around a room that you already know (if you could do this with your eyes open). These positive observations show that the problem is specifically with the visual system. It can also be useful to compare temporary experimental impairment of function with more permanent clinical cases. As with you when your eyes are closed, people who lack sight from birth can speak, read (provided the text is supplied in Braille), and do many other things, just as well as sighted people – and their navigation in total darkness is much better. The multitude of capacities that unsighted people have (and the new capacities they develop that sighted people usually don't have) places tight limits on the mental abilities that you can presume that the eyes directly support. It is clear that eyes are not required for hearing and feeling or, despite their normal importance to a sighted person, for navigation.

That eyes allow you to see may have seemed obvious. But in science, it never hurts to check the supposedly obvious. More importantly, the simple visual experiment you have just done has shown you what aspects of the mind are intact when vision is blocked. But, before we can treat this as 'case closed', we need to answer the mirror image question: Can we see (convert incoming light to images in our minds) without our eyes? Again, we should not jump to the 'obvious' conclusion without checking the available data. Let's look at the evidence for 'finger reading'.

Rosa Kuleshova could supposedly read newspapers with her fingertips. The Chicago Daily Tribune on Monday, 14 January 1963, has a story describing reports in the Russian government newspaper, *Izvestiya*, of Rosa's scientific testing by two separate institutes of the Russian Academy of Sciences:

> She was reported to have 10 light sensitive elements in every square millimetre of skin surface. The fingertips also were reported able to perceive red, blue, and green light in the same way as ordinary colour vision.... Several authorities at the National Institute of Neurological Diseases and Blindness said the Moscow report did not sound plausible. They knew of no similar case in the United States.

Results that depart from accepted theory require special scrutiny before we accept them. In this case, it needed a stage magician to provide appropriate scrutiny. (Scientists do not expect the universe to be fooling them nor have extensive experience of the mechanics of stage 'magic' tricks.) Sadly, there is no evidence that someone can read a newspaper with their fingers. People who seem to be reading text with their fingers turn out to be 'taking a peek' past the blindfold used or using some other form of deceit.[5]

But don't dismiss your fingers' sensory capacities. When I was an undergraduate student, we did an experiment (based on the Kuleshova reports, which were recent at the time) that showed that most of the students in the practical class could make a simple black/white judgement using only their fingertips. Oddly, I (as well as many of the other students) could not work out, straight away, how we were getting the correct answer – we just 'knew' what the answer was or sometimes, which was a bit more unnerving, thought we were guessing and found that we always guessed right. Try Experiment 2.2 if you want to check your own capacity for 'Dermo-optical perception' (you should be able to get correct answers) and test whether it depends on your fingers detecting light. You will find that seeing with light really does depend only on the eyes.

Experiment 2.2 Can you see with your fingers?

To check this out, you will need the help of a friend, a tea towel, and a piece of metal plate. Paint half of the surface of the plate white and half black. Sit at a table and arrange the tea towel so it hangs down over your face and tucks in under your chin. This is so you can't see the table – if you had a blindfold and wriggled your nose a bit, you could let enough light in to see the table with your eyes and read the paper, provided you hold your head at the right angle.

Get your friend to spin the metal plate around several times and then place it in front of you with one colour to the left and the other to the right. Hold your fingers over the surface – fairly close but not touching, so you can't detect textures – and guess which colour is to your left. Your friend should tell you whether you got this right or wrong, and then spin the plate again so it comes up with black or white to the left at random. With a little practice, you will find you can tell which colour is in which position most of the time – and you may not be able to tell how you are doing this! However, if you repeat the experiment in the dark (switching a light on only to check the result), you will find you still get the correct answer most of the time.

So, you are not using light, and therefore not using finger vision, to make your choice. Spinning the plate in the dark also rules out the possibility that you get the answer by telepathy from your friend. They will not know the correct answer, themselves, until you turn the light back on. But it does not rule out precognition. Now put a sheet of clear plastic over the top of the plate – your performance will drop to chance (i.e., 50% correct). This rules out light (again), telepathy (again), and precognition since none of them should have been affected by the plastic and your friend can still see the correct answer. So, what is the explanation? Clear plastic is opaque to heat waves, and it's actually the different amounts of heat reflected from and to your hand by the white and black parts of the plate and detected by temperature-sensitive cells in your fingers that is allowing you to make the correct choice.

After doing Experiments 2.1 and 2.2, you should have a good reason to think that your sight depends on light entering your eyes, and not your fingers, and that if you have a 'third eye',[6] it does not help you experience seeing very much. But, how much of vision depends on the eyes themselves and how much on other[7] parts of the brain? Where does our mental picture of the world come from?

The data suggest that we construct our visual 'world' from the evidence sent to the brain by our senses. This can give apparently contradictory results in different conditions. When you get to the end of this sentence, move your eyes to look away from the full stop and then back to the same full stop again. Your gaze will appear to have jumped across a stationary world (perhaps with a very brief jiggle).

Of course, the world doesn't usually appear to move when you change the point you are looking at. But the fact that it does not do so actually requires some explanation. To see why, cover one eye with one hand and *gently* press on the outside of the lid of the other eye to move the eye sideways while keeping it open. You see the world move! But the image will have moved at least as much across the retina when you moved your eye voluntarily. How can your experience of the world behave differently in the two cases? It turns out that your brain does some complex processing whenever you move your eyes to make what you see – your internal picture of the world – match the fact that the world hasn't actually moved.

When you move your point of gaze using your eye muscles, vision processing first receives a message from eye movement control telling it what is about to happen. Vision processing can then adjust your internal model of the world just the right amount in the opposite direction to cancel out the effect of the movement of the image – allowing you to see the world as stationary.

When you move your eye with your finger, there is no message to the brain about the upcoming eye movement. So, vision processing interprets the movement of the image on the retina as a result of the world moving (as in an earthquake or slab avalanche, for example), and so you see the world move by the same amount as the image did.

If you paralyzed your eye muscles (don't try this yourself!) and then tried to look to one side, you would see the world move in the opposite direction to your intended eye movement[8]: eye movement control would be telling vision processing to move the world to compensate for the eye movement but the shift in image position would not in fact occur because the eye muscles are paralyzed.[R13]

If you put these experiments together, you can draw two conclusions: (1) Light input to the eyes is necessary if we are to see the world. (2) But 'seeing', detecting objects in the world, needs a lot of neural processing.[9]

Anatomists have traced neural pathways from the eyes[10] into the brain (Figure 2.1). How does what we see change when these are damaged? The first major relay within the brain is the 'visual thalamus'.[11] If yours is destroyed, you cannot see; it is like losing your eyes. However, you will still be able to hear, smell, etc. – and use sunlight to help you adjust to jet lag!

Your sleep-wake cycle adjusts to seasonal changes and jet lag via a light-driven reset. So, to adjust to a new time zone, look at the sky – but never at the sun – regularly at the local time that you want to get up.

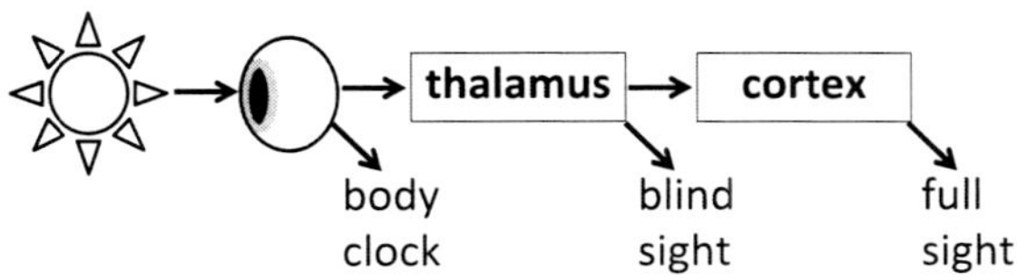

Figure 2.1 Parallel visual pathways. Light enters the eye and then goes via different pathways from the retina to a range of distinct brain areas that then produce changes in your body clock, blindsight, or full ordinary seeing. This figure shows just two pathways: one controlling the body clock and one going to the thalamus for fast processing and then on to the cortex for more detailed analysis. There are at least eight other areas that receive direct output from the retina, see Weiskrantz.[R31]

Surprisingly, this works even for someone whose visual thalamus is damaged. Here, as in many other cases, the brain is running parallel systems. The 'jet-lag' body clock has its own, separate, input from the eye.[R7] So, sunlight resets it even without object vision. It is also particularly sensitive to blue light. Do not use a computer monitor or phone late at night. This can cause sleep problems. If you are screen-addicted, wear glasses that filter out the blue light.[R19] Conversely, getting blue light from the sun in the morning helps to reset the clock.

Blocking the next stage of visual processing produces a particularly surprising effect. The visual thalamus is connected to the visual cortex[12] at the back of the head (Figure 2.1). Someone who has lost their visual cortex will insist that they cannot see anything. However, suppose you say:

> Look, I'm a psychologist. You know how we sometimes ask people to do crazy things. Just humour me. Whenever I ask, will you please just guess and press the left button if you might be looking at a square and the right one if you might be looking at a circle – but you must press one button or the other.

Although they insist that they cannot see and have to be told when to guess, they will have a high percentage of correct choices. They can also move their eyes accurately to a stimulus. This paradoxical capacity to make correct visually guided responses while insisting that you can't see anything is called 'blindsight'.[R31] Interestingly, this occurs not only with simple choice responses to simple stimuli but also with emotional reactions to facial expressions.[R15] So, your unconscious vision may control much more than you think it does.

To show blindsight to you, I can't ask you to temporarily block your visual cortex the way I asked you to block your eyes in Experiment 2.1. However, with Experiment 2.3, you can use a simple illusion[13] to show that you use blindsight all the time.

Experiment 2.3 Parallel visual systems

For this, you need a friend with a ruler, or a tape measure, or a camera on a tripod. Look at the middle dot of the cluster of dots on the left-hand side at the bottom of this box, and then close your eyes and reach for it with your first finger and thumb. Get your friend to check (with the ruler, tape, or by taking a picture) how far apart your thumb and first finger are as you do this. Next, do the same thing with the same hand for the middle dot in the cluster on the right-hand side. Your friend will find that your thumb and finger are much the same distance apart for both clusters. Now just sit back and look at the middle dot on the left side and then the middle dot on the right side. Do they look the same size? For most people, the one on the left looks much bigger than the one on the right. Finally, measure the middle dots with a ruler. They are the same size. So, your fingers got the size right (no, this is not finger-sight, it's eye-guided reaching) but your conscious vision, which you would use to report size verbally, got it wrong. What is going on here? The answer is that you used fast, lower-level, visual circuits (your blindsight) for non-conscious control of reaching, but you used slow, higher-level, illusion-prone, visual circuits for your conscious judgement of size. Only the higher-level visual circuits have an illusory perception of size. It's your highest-level thinking that is most mistaken!

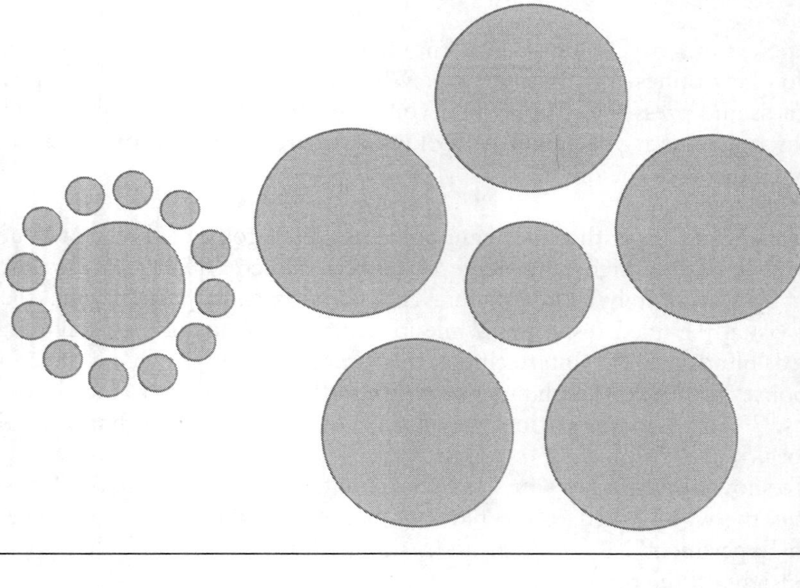

The bottom line is that you have both fast unconscious and slow conscious circuits. Processes like blindsight do not suddenly come to life only when the brain is damaged. The loss of a process through brain damage merely makes the number of processes particularly obvious. Blindsight is a mental faculty that is there all the time. But, of course, you never consciously know when you are using it – or any other unconscious process that contributes to your decision making.

A particularly useful way to look at these paradoxical results is that you have two different reasons for seeing: vision-for-action (guiding actions) and vision-for-perception (identifying objects).[R10] Vision-for-action needs to get the position in space right and so uses calculations that do not produce illusory sizes. Vision-for-perception needs to identify the nature of an object at any distance, and its calculations are better at doing this even if their size estimate is wrong.[14]

Cognition

That was an overview of simple perception, particularly seeing. How we detect objects; something an AI system might do fairly easily. Also, we saw that broken parts may still allow the brain to do things we might not expect.

But what about 'cognition'? Is *thinking* controlled by the brain? Does a fractured brain underlie the fractured mind[15] that we call schizophrenia?

Well, yes. How your simple visual perception works turned out not so simple. Schizophrenia is really complicated. It involves two quite different ways that brain dysfunction can affect the mind, and distorts multiple mental functions.

How does a fractured brain produce a fractured mind? This is way beyond the visual losses with a damaged visual system. Meet Gerald. I have transcribed a sample of him speaking below as best I can. But transcription loses nuances. You can get an idea of Gerald's speech; and how it goes with his mannerisms and how this all relates to the rather threatening situation he faces: a large group of doctors and other hospital people in authority over him. Just go to https://www.youtube.com/watch?v=gGnl8dqEoPQ.

Doctor: How are you doing?

Gerald: I don't feel so hot. I think I feel as though people have called me here to electrocute me, judge me, put me in jail, or kill me, electrocute me, because of some of the sins I've been in.

Doctor: Is this a new feeling for you?

Gerald: – [pause] – the main thing is don't get excited but the thing is, is ungh, ... it's not a new feeling, no. – [slight pause] – I'm scared of people.

Doctor: This must be very frightening for you then.

Gerald: I'm so scared I could tell you that picture's got a headache.

Doctor: Can you tell me more about that "the picture has a headache"?

Gerald: You wanna know?
Doctor: Yes, I do.
Gerald: OK, when a sperm and an egg go together to make a baby, only one sperm goes up in the – [slight pause] – egg and when they touch there's two contact points that touch before any other two and – em – its carried up into the egg and when they fuse its [speech becoming faster and continuous] like nuclear fusion except its human fusion there's a mass loss of the proton one heat abstraction goes up in the electron spins around comes back down in the proton to form the mind and the mind can be reduced to one atom ...

Gerald's loose connection of thoughts and sometimes bizarre associations are typical of active schizophrenia. It's more obvious in the YouTube video that additional symptoms include unusual mannerisms (Gerald keeps twirling a chunk of hair on the top of his head), hallucinations, delusions, and paranoia ("I feel as though people have called me here to electrocute me"). These are all called 'positive' symptoms. We think they are linked to excess neural activity; and (luckily) sensitive to antipsychotic medication. Augustine, off and then on effective medication, gives you an idea https://www.youtube.com/watch?v=vRTH3VrdVDU.

Schizophrenia also involves blunted emotional responses and a general poor quality of speech. These 'negative' symptoms are believed to be linked to insufficient neural activity, and are not sensitive to antipsychotic medication. This is most obvious with Augustine after treatment. But you can see signs of them in Gerald, too.

The positive symptoms of schizophrenia, relieved by drugs, are thought to be the result of excess brain activity. The negative symptoms are linked to brain areas that are smaller than in healthy people. Damage to these disturbs the same mental functions as does schizophrenia: prefrontal cortex (planning/thought); temporal cortex and hippocampus (memory); basal ganglia/striatum (motor control); and thalamus (which we already met in the context of sensory input but which is also involved in the control of memory). This is not a simple set of structures and damage to them is variable.[16]

As with vision, then thinking, memory, and speech depend on activity in your brain. As with blindsight, the brain fracture of a schizophrenic illustrates your own capacities. Mirroring the reduction of schizophrenic hallucinations by therapeutic dopamine antagonists, hallucinations can be induced in anyone with drugs such as LSD, mescaline (in peyote cactus), psilocybin (in magic mushrooms), and PCP (phencyclidine). You may already have personal evidence of this. But do not attempt to test it directly. Not only are these drugs illegal in many countries but even a single dose can have long-term negative effects.[R1]

There should be enough anecdotal evidence to convince you that altering your brain *would* alter your thinking. Interestingly, this altered thinking can cause "complete mystical[/religious] experiences in the majority of people studied".[R4] Excitingly,[17] low doses of psychedelics show improvement in a wide range of psychiatric disorders. A single dose is often immediately effective as an adjunct to psychotherapy.[R26] But, definitely see your doctor rather than trying to self-medicate – things like exact dose and appropriate support can be crucial for a positive effect as opposed to a 'bad trip'.

Emotions and feelings

Contrasting with cold calculating cognition, and most important for our everyday well-being, are feelings and emotions. These are strongly linked to 'mental disorders'. Work days lost to mental disorders are double those lost to other health issues.[R27] With mental disorders in general, "the economic value associated with this burden is estimated at about US$5 trillion. At a regional level, the losses could account for between 4% of gross domestic product in Eastern sub-Saharan Africa and 8% in High-income North America".[R2, p. 1]

Disorders of anxiety and depression, in particular, are also important for the health budgets of developed countries, with about one person in five suffering from a mood disorder. Anxiety disorders are currently the most prevalent psychiatric diseases in Europe and the United States,[R18] the sixth highest cause of years of life lived with disability,[R5] and so a serious and increasing cost for health care systems.[R16,17] In 2010, anxiety disorders were linked to 7% of all suicide mortality globally.[R6]

With schizophrenia, I emphasized its strong neurological component – the physical differences in the brains of schizophrenics, which appear to have their origin in early brain development.[R28] In contrast, it appears possible to trigger mood disorders such as post-traumatic stress disorder in anyone if they are exposed to sufficient (particularly chronic) stress.[18] The states of anxiety and depression are also quite different from, for example, the hallucinations of schizophrenia in that they are adaptive and occur frequently, if briefly, in everyone. Anxiety and depression are only thought of as being disordered when they are excessive, chronic, or inappropriate.

Negative mood states such as anxiety evolved to keep us out of trouble. Although it is often unpleasant, anxiety – like pain – is important if you are to assess risk, survive and reproduce. The dangers of insufficient anxiety are reflected in traffic laws that limit the use of one of the oldest anti-anxiety drugs – alcohol. High doses can result in poor coordination and attention. But the main danger of alcohol is that even lower doses reduce the driver's perception of risk.

To convince yourself that mood depends on the brain, you need to do no more than drink a standard glass of an alcoholic beverage or smoke one cannabis cigarette (provided either is legal at your current location and you are over age). You can also produce somewhat opposite effects with a standard shot of coffee. But, contrary to popular belief, coffee does not reverse the impairment of driving by alcohol; it just masks the more personally obvious symptoms https://www.cdc.gov/alcohol/fact-sheets/caffeine-and-alcohol.htm.

But chronic negative mood is maladaptive. When someone has unremitting severe depression, for example, they may regularly think seriously about committing suicide. It is a common experience nowadays that people with *severe* depression will often be helped by taking an antidepressant drug.[19] Conventional antidepressants[20] can not only reduce depression, but they can also reduce panic attacks, anxiety, and often obsessionality.

However, therapeutic improvement is very slow. The depression can reduce steadily over weeks, months or even years. But the effects of standard antidepressant drugs on the key brain chemicals, serotonin and noradrenalin, are immediate. So, the drugs[21] are affecting brain processes that gradually change what we can best see as a patient's personality.

Whether you are generally more/less anxious or more depressed/manic are only two of many aspects of your personality that are controlled by your brain. For many people, it is these longer-term, systematic, characteristics of the mind that they would most see as defining their 'self'.

Personality

The last mental characteristic I want to look at in brain terms is personality. By definition, this involves differences between people that are stable over long periods. Many personality trait labels (creativity, assertiveness, gender) are not clear. Where they are *defined* via self-report (including lexically derived scales), we do not know if they link to biology.

To make at least one biological link, I will focus on sexual orientation. This is externally observable (except in some societies where particular ones are proscribed). This relates to 'straight/gay/other' and 'male/female' – but only if you use these words as distinct from gender identity. External body form is one thing; preferred type of sexual partner another; and gender identity different, again. An important point about gender in the sense of personal perception or social construction is that it goes way beyond simple reproductive behaviour. Critically, unlike sexual orientation, we cannot determine gender orientation in the rats that are our best source of mechanistic biopsychological data – because we cannot ask them.

Basic biological processes determine both sex in terms of external body form and sexual orientation – the latter through internal brain systems. Even the most concrete of these attributes is not conventionally male/female. Multiple factors determine these attributes and are graded rather than categorical. But your specific body type and sexual orientation (in the limited sense above) were largely determined very early in development (and crystallized further during puberty). Changing them is hard (or, in the case of things like male/female, X/Y, chromosomes, impossible). I hope you will find this melting-pot alternative to a binary world of identical clones interesting and illuminating rather than disturbing.

Sex and the brain – let us start with Simon LeVay [https://en.wikipedia.org/wiki/Simon_LeVay; https://www.simonlevay.com/my-resume]. LeVay is a gay neuroscientist. In a sudden change of research direction, he set out to see if there was evidence that his sexual orientation was 'wired in'. He was working in California at the time when many gay males started dying of AIDS, including his long-time partner. He analysed brains *post mortem*, comparing those of males and females. As he had expected, he found brain differences. In particular, a brain nucleus, INAH3,[22] tended to be the same size in gay males as straight females and smaller than the INAH3 of straight males.

The simple observation that gay males have a size of INAH3 typical of straight females in a male body requires scrutiny. Most of LeVay's 'male gay' brains had died of AIDS. But we can rule out AIDS as the cause of the reduced INAH3. Some of the straight males died of AIDS with a 'straight' INAH3.

Importantly, later brain imaging found[R29] that gay males without AIDS[R29] were like straight females – with more widespread connections from the *left* than the right amygdala to one set of shared target structures. In parallel, gay females and straight males have similar widespread connections from the *right* than the left amygdala to a second set of shared target structures. As with LeVay's results, this suggests that at least some parts of the brains of gay people are similar to those of straight people of the opposite sex, but it also makes clear that INAH3 is not the sole example.

But other 'gay brain' differences do not show such a simple 'opposite sex brain' pattern:

> [Gay] orientation seems associated with a less pronounced sexual differentiation of white matter tracts and a less pronounced functional connectivity of the self-referential networks compared to straight orientation. Analyses of [cortical thickness] suggest that male and female [who are gay] are not simple analogues of each other and that differences from straight controls are more pronounced in [gay males].
>
> (LeVay[R23, p. 1175])

Interestingly, gender dysphoria is linked to quite separate self-body perception networks.[R22,24] Thus, sexual orientation and perception of one's own preferred sex are distinct, providing a complex background for any social construction (which is likely weak[R32]).

When and how is brain sexual orientation determined? In studies, with 'identical' twins recruited from twin registries[23] (who share similar developmental environments, the same uterine environment, and the same genes), if one is gay, there is about a 25% chance that the other will be. In contrast, with non-identical twins (who share similar developmental environments, the same uterine environment, but only half their genes), it is 15%.[R3] Such 'concordance' rates can easily be compared with each other, but their absolute values can depend on environmental factors and measurement error, so you don't want to take the specific percentages too seriously. Many studies use self-report of gay orientation, which may also bias to low reported rates. The key take-home message from these genetic effect *differences* is that there is *some* genetic control of sexual orientation that, together with uterine influences, produces effects earlier than could be accounted for by social and other learning-related influences. So, how are these early effects produced?

As background to how your sexual orientation is determined, let's first look at what happens with the sex of your other body parts. In the absence of hormones, the basic body masterplan of a developing foetus is female. Switching on certain genes before you are born produces hormones that change your body along masculine lines. Some hormones (such as Müllerian Inhibiting Substance) 'defeminize', they *remove female* characteristics. Other hormones (such as testosterone) 'masculinize', they *add male* characteristics. A flood of hormones at puberty completes the job. The flood then subsides.

But you can see how a body reacts to such hormones, even later, from the way artificial versions of male hormones are used to enhance athletic performance in males and females; from the fact that therapeutic uses of testosterone (T)[24] in peri-menopausal women can increase femininity and mood stabilization at low doses and can produce male pattern hair growth at higher doses; and from breast development in males given T blockers to reduce prostate size. Similar, relatively short-term effects are seen with, for example, the fact that "traders made more money on days when their testosterone levels were highest".[R14, p. 5]

Your brain develops sex-linked characteristics the same way as your body. This shouldn't be too surprising, as the brain is just another part of the body. Hormones during development make the size of some brain areas more male-typical or female-typical. The effects of hormones on the brain have been most clearly demonstrated in experiments with rats: removing neonatal T gives a male rat a female-sized medial pre-optic nucleus, while administering T gives a female rat a more male-sized medial pre-optic nucleus.

Surprisingly, brain sex is not determined at the same stage of development as body sex (Figure 2.2).[R11] A genetically female rat will have a female body plan at birth, but, if injected with T within 4 days of birth, it will develop a more male-sized medial pre-optic area without changing their body plan. Conversely, a genetically male rat lacking T from day 1 after birth (but not if the lack is from day 21 on) will have a female-sized sexually dimorphic nucleus but a male body. These changes fit with the patterns we already saw in the adult human brain. Additional changes occur at puberty. While this 'simple' T story may seem to you to have already produced a large array of possible results, it should be noted that even if we look only at typical male (mounting) and female (lordosis) mating behaviour in the adult rat the results do not depend simply on peri-natal and peri-pubertal T but also on its interaction with, for example, the hormone kisspeptin[R25] to produce masculinization and defeminization.

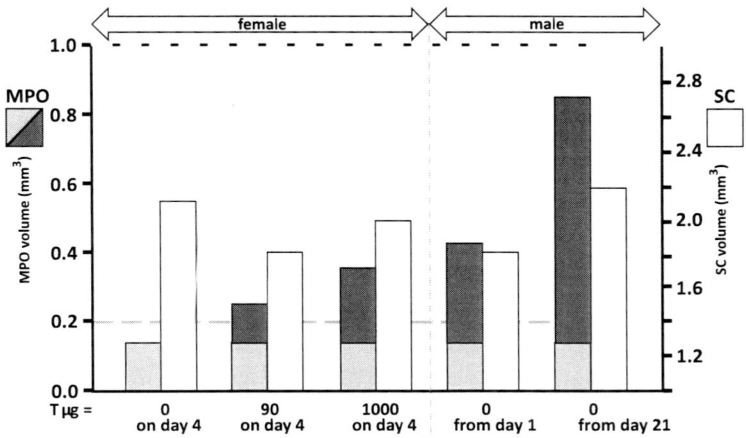

Figure 2.2 Timing of brain sex changes. Effect on the size of the adult medial pre-optic nucleus (MPO) and suprachiasmatic nucleus (SC, white bars and right-hand scale) of nominal dose of testosterone (T) delivered on the stated day after birth in female (day 4) or eliminated from male rats (from day 1 or day 21). The dashed lines represent the size of the MPO in untreated females (0.2 mm^3) and males (1.0 mm^3). Light grey bars, size of female oil-treated control MPO (repeated across other conditions for comparison). Dark grey bars, increase relative to oil-control female MPO produced by 90 or 1,000 g of T in females and by pre-birth natural T (from day 1) or post-birth natural T (from day 21) in males. Redrawn and adapted from Gorski et al.[R11, Fig. 5]

Unfortunately, I don't have a measure, manipulation, or drug that I could suggest to show you personally that your sexual orientation is controlled, at least in part, by your brain's anatomy. One hope for such is the use of brain imaging (although this is costly) – but at present this does not seem to be giving clear answers.[R8] However, both the rat and human brain data suggest that key aspects of sexual orientation are set in the brain very early on and are difficult or impossible to change. So, instead, I'll ask you simply to imagine changing your sexual orientation. As with disliking blue cheese or being unable to play professional-level golf, "the reality is that [gay people] have no choice – we are born that way and no amount of hectoring or hypnosis can make us change" (Sir Elton John, Otago Daily Times, July 2000, see also http://news.bbc.co.uk/2/hi/entertainment/792128.stm).

We like to put things into neat little boxes. When your body is masculinized and/or defeminized this is usually obvious, with clear positive or negative physical changes relative to the default body plan. Most frequently masculinization and defeminization occur together, and the result is a 'male'. But, given that they can occur independently, there are four nominal basic types for a body (Figure 2.3, rows). Given the difference in timing, there are four basic types for a brain (Figure 2.3, columns). The

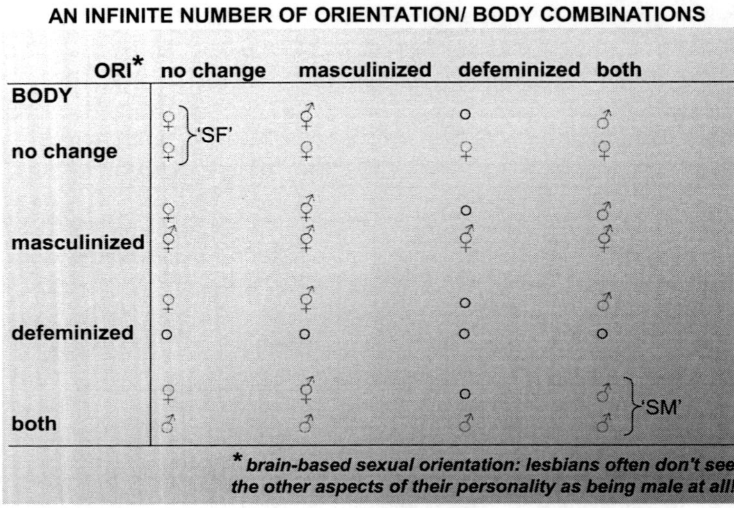

Figure 2.3 Infinite variation in body form and sexual orientation. Your brain and body can be independently masculinized and/or defeminized or not. In categorical terms, this would deliver 16 body/orientation combinations (as illustrated by the symbols; ♂ = masculinized; ♀ = not defeminized; o = defeminized but not masculinized). This ranges from neither masculinized nor defeminized (i.e., straight female, SF) to both brain and body being both masculinized and defeminized (i.e., straight male, SM). But hormonal changes are graded (like the grey background but in at least four dimensions). So, there are really an infinite number of orientations occupying a dimensionally complex data space.

result is a lattice of 16 nominal brain-body type combinations. So, if the changes occurred in a categorical fashion, there would be at least 16 combinations of body form and sexual orientation. But there are no clear dividing lines. For example, T (producing masculinization) can vary in amount at the critical times. It is not simply there or not there. As an extra source of variety, sexual orientation is not strongly linked to other aspects of male/female personality differences:

> The ultimate challenge will be to establish how the genetic differences among individuals interact with environmental factors to produce the diversity that exists among us. But to approach this goal, it will be necessary first to reject the notion of the tabula rasa – the idea, propounded by Locke, that the mind of the newborn child is a blank slate on which experience is free to write whatever it will. In reality, our range of individual development is defined and limited by what we are born with. Like waterlilies, we swing to and fro with the currents of life, yet our roots moor us each to our own spot on the river's floor.
>
> (Le Vay[R21, p. 138])

What does chimpanzee love tell us about human sex?

Before moving on, it is worth noting that the insights that our brain processes give us about our minds can be greatly increased by asking why a particular process occurs. For example, in talking about sexual orientation, gender, and related issues, we have to ask why people 'have sex' and whether 'making love' is just a euphemism. If having sex were just to produce children, it would only be necessary for a brief period every 2–3 years.[R30]

Here, comparison among species can provide clues. Since I live in Dunedin, a short drive from the only mainland colony of Royal Albatross in the world, I'll pick albatross as one example species and bonobos (thought to be our closest relative) as the other. Let's ask how and why they have sex. The data comparison is in Table 2.1. Put briefly, albatross do what would be expected if sex occurs only for reproduction – but with extensive bonding rituals to maintain continuous two-parent support for their one chick until it is independent. They both take a rest year and reconnect with bonding rituals like beak clashing when they return.

Bonobos, in contrast, have a 5–6-year breeding period with child care filling the gap years. But, like humans this is not linked to either the frequency or the way they 'indulge' in sex:

> For these animals, sexual behaviour is indistinguishable from social behaviour. Given its peacemaking and appeasement functions, it is not surprising that sex among bonobos occurs in so many partner combinations, including between juveniles and adults. The need for peaceful co-existence is obviously not restricted to adult heterosexual pairs.
>
> (de Waal[R6, p. 86])

Table 2.1 A comparison of various characters of sexual and related behaviour between albatross and bonobos

Character	*Albatross*	*Bonobo*[R6]
Reproduction/female	Every 2 years	Every 5–6 years
Sex frequency	One period per 2 years	Frequent
Female receptivity	One period per 2 years	Almost continuous
Sex position	Ventral	Dorsal or ventral
Sex partners	One, almost always opposite sex	Various, same or opposite sex (appears to include female:female orgasm)
Sex variations	None	Genito-genital rubbing (various); oral sex; massage; tongue kissing
Sex as conciliation	No	Sex or variation either after or instead of aggression 'make love not war'
Economic sex	No	Sex for food
Other bonding	*Yes*	*Yes (extensive rituals)*

That said, it is important to note that bonobos also have extensive non-sexual bonding rituals (consistent with them being much more social than albatross). Thus, in humans, we can see 'lust' as existing to produce offspring, 'love' existing to support albatross-like pair bonds for the support of offspring (and the offspring of offspring) while they are developing, and social bonding as existing to improve more general survival and so reproductive capacity.

Progressing to mental mechanics

I hope I have now convinced you that your mind depends fundamentally on your brain. Neural activity is mental activity. Neural processing is cognition. Long-term neural sensitivities produce mood, and more consistent aspects of neural reactivity produce our personality. These all work together in the context of the selection pressures that drive evolution. The brain gives us a key to the mind. The fundamental mechanisms are *Mental Mechanics*.

At the end of this book, I will try and convince you that understanding the mind's mechanisms can lead to understanding the mind itself[25] and its disorders (Chapter 10).[26] But first, I will use data from other species and other people, particularly people with temporary or permanent brain dysfunction, to derive specific conclusions about how *your* mind works. I will also continue as far as possible to give you tests you can apply to yourself or your friends that will allow you to check what I am telling you about your brain directly via your own mental experience.

But interpreting the effects of brain manipulations on the mind requires some care. The cause of a reported or observed mental change may not be obvious (like heat detection in the finger vision experiment). It's also at least as important to check what aspects of the mind were not affected (Experiment 2.4). This checking of alternative explanations can be complicated and hard work, and so it's often tempting to take your immediate conclusions from a particular result at face value – especially if you want to make newspaper headlines (fake or otherwise).

So, when next you read a sensational article about brain function, your first step should be to check that obvious alternatives have been ruled out. In what follows, I will usually not describe all the necessary checking of alternatives, but if, for example, I explicitly link vision to a particular brain area, then you should assume that appropriate experiments have been done to rule out hearing.

We will analyse mental basics without having to solve the problems of the mind as a whole. To get a picture of the mind, we need to tie together different findings about different parts of the brain. To help with this, we can call on a large body of data on how the brain is built. How do the brain's processing units arise? How are the thousands of millions of these units in the brain organized? The next two chapters deal with these intertwined questions, providing a detailed bedrock for understanding the mind. But if you prefer a top-down to bottom-up approach, you may prefer to jump to Chapter 5 now and return to Chapters 3 and 4 when you feel a need for their level of detail.

Experiment 2.4 Testing a spider's hearing?

There is an old joke about a professor who trains a spider to move on command and brings it as a demonstration to class. 'Walk', says the professor; and the spider walks. 'Stop', says the professor, and the spider stops. Then the professor pulls the spider's legs off and says 'walk' and the spider doesn't move. "This demonstrates that it hears through its legs", says the professor to the class. Unfortunately, this type of conclusion is seldom an obvious joke for brain scientists (try entering 'hearing legs' in Google). It is often very hard to be sure you have ruled out alternative explanations (see Experiment 2.1). To prove this particular conclusion, the professor would also need to have trained the spider to walk when a light was flashed. If it doesn't walk on verbal command and does walk when the light flashes, then you have evidence that it hears through its legs – but, if it does not walk under either condition, you have clearly not just affected its hearing.

Notes

1 I'll leave until the last chapter questions about what the mind is, and related questions about consciousness. I have emphasized 'mechanical' here because deciding about the presence or absence of a mind in the hero of "I, Robot", who has a positronic brain, would present a significantly different problem from current digital systems, even when the current systems can pass the Turing test. Those digital systems can be treated as being, in principle, just like the early cogs-and-wheels 'Babbage Difference Engine'.
2 Detailed in Chapters 3 and 4.
3 In Chapter 6, we will use a simple mollusc, the Sea Hare to look at the different types of learning that can change your behaviour.
4 An exception to this is when the senses get entangled in a condition called synaesthesia. See https://en.wikipedia.org/wiki/Synesthesia.
5 See also 'Dermo-optical perception' in Wikipedia.
6 A mystical concept occasionally linked to the pineal gland – see 'Third Eye' in Wikipedia.
7 The retina of the eye has a lot of complex neural circuitry and can be treated as being part of the brain.
8 Since this is a subjective report of experience, we should note that there was, obviously, no way of controlling for knowledge of the paralysis with this experiment. The report I have cited is also rather indirect "The eyes are directed by judgment (based on past visual experience) of the amount of effort needed to move them. If the relationship between effort and resultant movement is grossly disturbed, for example by curare, the position of external objects is misjudged and eye movements are accompanied by apparent movement of the external world, as one of us (P.M.) has recently confirmed." So, the participant was one of the paper's authors.

9 We'll look at this in more detail in Chapter 9.
10 The retina has a lot of complex neural circuitry and so can be thought of as an initial part of the brain – this is also true of the olfactory bulb.
11 The thalamus processes a large proportion of incoming sensory information and distributes this information to, for example, the primary sensory cortex, such as the visual cortex, as shown in Figure 2.1. However, it is also involved in re-circulating information in support of functions such as memory, so it does more than pass on sensory input.
12 Technically, the primary visual cortex, striate cortex or Area V1. For more details on visual processing, see Chapter 9.
13 This illusion is called the Ebbinghaus illusion. The demonstration that the fingers don't suffer from it actually involved using poker chips in place of the central circles and had a variety of control procedures.[R12]
14 See Chapter 9.
15 The name's roots are: schiz (skhizein (σχίζειν), to split) = i.e., shattered, not multiple; and phren (φρήν) = mind. It is quite different from multiple personality disorder.
16 We return to this and the circuits involved in Chapter 10 *Schizophrenia – a different 'double' hit*. Neural involvement can also be seen at a finer grain level. Recent automated, AI, analysis of semantic trajectories in a category- and letter-verbal fluency task found that "the influence of semantic similarity on behavior was reduced in schizophrenia relative to controls, predicted negative psychotic symptoms, and correlated with a MEG [MagnetoEncephalography] signature of ['trait-like'] hippocampal ripple power".
17 See also Chapter 10.
18 See "Nature and nurture – events" in Chapter 10.
19 With mild depression, improvements in mood are often due to psychological 'placebo' effects rather than the pharmacological action of the drugs.[R20]
20 Common examples are fluoxetine (Prozac) and paroxetine (Aropax), which are members of a class of antidepressants known as specific serotonin reuptake inhibitors (SSRIs). These increase the availability of the neurotransmitter serotonin. A range of other classes of drug acting on the neurotransmitter noradrenaline or on dopamine, noradrenalin and serotonin can also treat depression. The different drugs within a class and the different classes differ mainly in their side effects.
21 Given my emphasis on drugs, here, it's worth noting that psychological therapies can also be effective – changing similar brain systems to the drugs. Also altering habits, attitudes, diet, and life circumstances to allow a drug-free life should be the main goal of therapy. Like aspirin, antidepressants may reduce the pain and allow you to proceed more or less as normal – but it's better to solve the underlying causal problems.
22 Interstitial **N**ucleus of the **A**nterior **H**ypothalamus, #3 – but **INAH3** is easier.
23 Targeted sampling gets higher values with 52% concordance for identical twins and 17% for non-identical.[R3]
24 Testosterone occurs normally in women but at lower than male levels, while "the most abundant active sex steroid in women throughout the female lifespan" being about ten times more than estradiol.[R9, p. 231]
25 Chapter 9.
26 Chapter 10.

References

R1. Abraham, HD & Aldridge, AM, Adverse consequences of lysergic acid diethylamide. *Addiction*, 1993. 88: 1327–34.

R2. Arias, D, Saxena, S, & Verguet, S, Quantifying the global burden of mental disorders and their economic value. *EClinicalMedicine*, 2022. **54**: 101675.
R3. Bailey, JM, Vasey, PL, Diamond, LM, Breedlove, SM, Vilain, E, & Epprecht, M, Sexual orientation, controversy, and science. *Psychological Science in the Public Interest*, 2016. **17**: 45–101.
R4. Barrett, FS & Griffiths, RR, Classic hallucinogens and mystical experiences: Phenomenology and neural correlates. *Current Topics in Behavioral Neurosciences*, 2018. **36**: 393–430.
R5. Baxter, AJ, Vos, T, Scott, KM, Ferrari, AJ, & Whiteford, HA, The global burden of anxiety disorders in 2010. *Psychological Medicine*, 2014. **44**: 2363–74.
R6. de Waal, FB, Bonobo sex and society. *Scientific American*, 1995. **272**: 58–65.
R7. Dibner, C, Schibler, U, & Albrecht, U, The mammalian circadian timing system: Organization and coordination of central and peripheral clocks. *Annual Review of Physiology*, 2010. **72**: 517–49.
R8. Frigerio, A, Ballerini, L, & Valdes Hernandez, M, Structural, functional, and metabolic brain differences as a function of gender identity or sexual orientation: A systematic review of the human neuroimaging literature. *Archives of Sexual Behavior*, 2021. **50**: 3329–52.
R9. Glaser, R & Dimitrakakis, C, Testosterone therapy in women: Myths and misconceptions. *Maturitas*, 2013. **74**: 230–4.
R10. Goodale, MA, Transforming vision into action. *Vision Research*, 2011. **51**: 1567–87.
R11. Gorski, RA, Gordon, JH, Shryne, JE, & Southam, AM, Evidence for a morphological sex difference within the medial preoptic area of the rat brain. *Brain Research*, 1978. **148**: 333–46.
R12. Haffenden, AM & Goodale, MA, The effect of pictorial illusion on prehension and perception. *Journal of Cognitive Neuroscience*, 1998. **10**: 122–36.
R13. Hammond, PH, Merton, PA, & Sutton, GG, Nervous gradation of muscular contraction. *British Medical Bulletin*, 1956. **12**: 214–8.
R14. Herbert, J, Testosterone, cortisol and financial risk-taking. *Frontiers in Behavioral Neuroscience*, 2018. **12**.
R15. Heywood, CA & Kentridge, RW, Affective blindsight? *Trends in Cognitive Sciences*, 2000. **4**: 125–6.
R16. John F, Cryan, JF, & Sweeney, FF, The age of anxiety: Rrole of animal models of anxiolytic action in drug discovery. *British Journal of Pharmacology*, 2011. **164**.
R17. Kessler, RC, The global burden of anxiety and mood disorders: Putting ESEMeD findings into perspective. *Journal of Clinical Psychiatry*, 2007. **68**: 10–9.
R18. Kessler, RC, Chiu, WT, Demler, O, Merikangas, KR, & Walters, EE, Prevalence, severity, and comorbidity of 12-month DSM-IV disorders in the National Comorbidity Survey Replication. *Archives of General Psychiatry*, 2005. **62**: 617–27.
R19. Kimberly, B & James RP, Amber lenses to block blue light and improve sleep: A randomized trial. *Chronobiology International*, 2009. **26**: 1602–12.
R20. Kirsch, I, Placebo effect in the treatment of depression and anxiety. *Frontiers in Psychiatry*, 2019. **10**: 407.
R21. Le Vay, S, *The sexual brain*. 1993, Cambridge: MIT Press.
R22. Majid, DSA, Burke, SM, Manzouri, A, Moody, TD, Dhejne, C, Feusner, JD, & Savic, I, Neural systems for own-body processing align with gender identity rather than birth-assigned sex. *Cereb Cortex*, 2020. **30**: 2897–909.

R23. Manzouri, A & Savic, I, Cerebral sex dimorphism and sexual orientation. *Human Brain Mapping*, 2018. **39**: 1175–86.

R24. Manzouri, A & Savic, I, Possible neurobiological underpinnings of homosexuality and gender dysphoria. *Cerebral Cortex*, 2019. **29**: 2084–101.

R25. Nakamura, S, Uenoyama, Y, Ikegami, K, Dai, M, Watanabe, Y, Takahashi, C, Hirabayashi, M, Tsukamura, H, & Maeda, KI, Neonatal kisspeptin is steroid-independently required for defeminisation and peripubertal kisspeptin-induced testosterone is required for masculinisation of the brain: A behavioural study using Kiss1 knockout rats. *Journal of Neuroendocrinology*, 2016. **28**: e12409.

R26. Nutt, D, Erritzoe, D, & Carhart-Harris, R, Psychedelic psychiatry's brave new world. *Cell*, 2020. **181**: 24–8.

R27. Oakley-Browne, MA, Wells, JE, & Scott, KM, *Te Rau Hinengaro: The New Zealand Mental Health Survey*. 2006, Wellington: Ministry of Health.

R28. Rapoport, JL, Giedd, JN, & Gogtay, N, Neurodevelopmental model of schizophrenia: Update 2012. *Molecular Psychiatry*, 2012. **17**: 1228–38.

R29. Savic, I & Lindström, P, PET and MRI show differences in cerebral asymmetry and functional connectivity between homo- and heterosexual subjects. *PNAS*, 2008. **105**: 9403–8.

R30. Waynforth, D, Ancestral birth spacing, in *Encyclopedia of evolutionary psychological science*, TK Shackelford and VA Weekes-Shackelford, Editors. 2021, Cham: Springer International Publishing. p. 285–9.

R31. Weiskrantz, L, Blindsight revisited. *Current Opinion in Neurobiology*, 1996. **6**: 215–20.

R32. Xu, Y, Norton, S, & Rahman, Q, A longitudinal birth cohort study of early life conditions, psychosocial factors, and emerging adolescent sexual orientation. *Developmental Psychobiology*, 2020. **62**: 5–20.

3 The mental processing unit

The neuron as a mental processor

Neurons are the key processing units of your mind machine.[1] They are like other cells (which are often 'excitable' and communicate with each other), but more so.

Importantly, there are a lot of neurons. The human brain is estimated to have 86 billion neurons.[R1] So, even if the neuron is a simple cell, your mind machine can do hugely complex things.

Helpfully, large groups of neurons follow rules. We can look at each group as a whole and understand *how* the neurons work in principle. More neurons just mean finer detail processing.

For example, there are about 125 million retinal photoreceptors in your eye. They basically all work the same way – like each pixel of a 125-megapixel digital camera. Each photoreceptor deals with one point in visual space, with the overall image fitted into a simple 2D array connected to the next processing level. Divide 86 billion total brain neurons by 125 million visual elements, assume other systems have the same properties, and the problem is already reduced to only 8,000 computational lumps we need to worry about in detail. Computer models can handle those kinds of numbers easily.

Chapter 3 explains how, as brains increased in size, they got more complex. But this growth followed some basic rules that can give you a simple picture of neuronal processing. Then you just say something like "and repeat this unit 125 million times" and you can understand the system. You will meet a lot of diagrams later where one neuron is drawn to represent thousands and that is all you need to understand how the large-scale circuits work.

But what about the *elements* of the circuits, the neurons? Unlike other cells, they have input trees ('dendrites'; Greek, dendron = tree) and output fibres ('axons'). Your axons can be over a meter long when connecting your spinal cord to your big toe using just one neuron.

The dendrites receive masses of connections that can generate electrical currents by opening channels. The tree gathers the currents, which flow down to the cell body. The cell body integrates the incoming information and can then send output to other neurons via its axon and their dendrites.

DOI: 10.4324/9781003465027-3

From a large-scale perspective, the brain can seem analogous to a computer. But its processing is quite different at the unit-by-unit level. Neurons are 'wetware' not 'hardware' + 'software'. Cell bodies act a bit like transistors in a computer, but are also like batteries. Neurons communicate in much the same way as the body cells of a sponge![2]

To understand this in detail, read on. If you are not keen on physics and chemistry, feel free to skip to the next chapter.

The cell is an ion prison and a battery

Cells, as the name suggests, have walls that allow the inside to be different from its environment. From the point of view of a potassium ion[3] (K+) the cell is a prison (Figure 3.1) from which it can occasionally escape through a door (usually referred to as a channel). From the point of view of sodium (Na+) and chloride (Cl–) that are mostly outside the wall, the cell is a place that they can only enter through a special visitor's door when they are permitted. There are many other ions outside and inside (including a variety of anions, A–).

With the doors in their normal closed position (it's a prison), a biochemical pump clears Na+ out of the cell while pulling K+ in. The key to the excitability of cells in general and neurons in particular is that, if there are a lot of, for example, Na+ ions outside, and only a few inside, the cell is like a battery. Think of the inside as one terminal of the battery (negative) and the outside as the other (positive).

In simple terms, a channel is just a hole through the cell membrane. It can be closed when no ions will pass through it. Or, it can be open – in which case it will make a circuit with ions passing through it, moving from areas of high concentration to areas of low concentration (see Experiment 3.1).

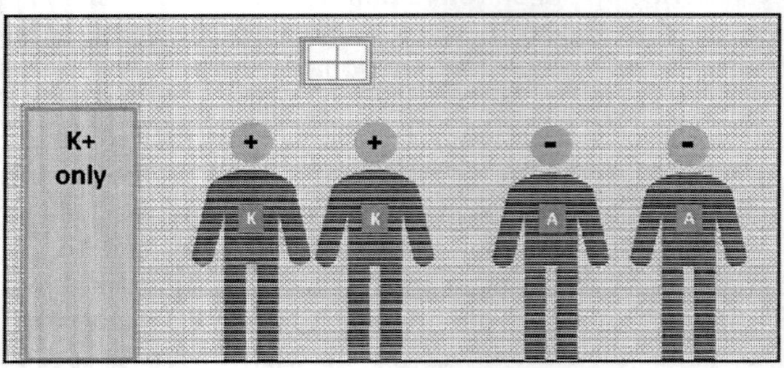

Figure 3.1 Cells and ions. Excitable cells have walls (lipid membranes) that prevent ions (the elements that make up molecules) from moving between the cell and its environment. This allows the amounts of specific ions to be different on the inside than outside. Here, potassium ions (K+) and anions in general (A-) are shown as prisoners. Sodium ions (Na+) and chloride ions (Cl–) are on the other side of the wall. K+ can escape if a channel through the membrane (acting like a cell door) is opened. Likewise, Na+ and Cl– can come in to visit. However, more like immigration queues than a prison, each type of channel will let through one type of ion but not another.

Experiment 3.1 How does diffusion work?

Ions diffuse through a channel like water moving down the spillway of a dam. The water moves because of a gravity gradient (high to low). Ions (or any molecule diffusing through any medium) move down their specific concentration gradient (e.g., high Na+ to low Na+). In this experiment, you will see that diffusion is not a result of gravity, that concentration is important, and (surprisingly) that the visually slow diffusion you observe at the macroscopic scale is at a rate that can produce very quick responses at the neuronal scale. (Neurons are very small, and their cell walls and channels are even smaller.)

For the experiment, you will need bottles of two different food colourings (ideally one red and one green, if you don't have these already, they should be available in any local food store). You will also need Sellotape/Scotch Tape (i.e., sticky, clear plastic tape, ideally at least 20 mm wide), water in a mug, a small bore (~5 mm dia.) drinking straw, some plain white paper, scissors, a teaspoon and two larger spoons (table or serving will do). Also, you will need a stopwatch (there should be one in your cell phone clock app) and a ruler.

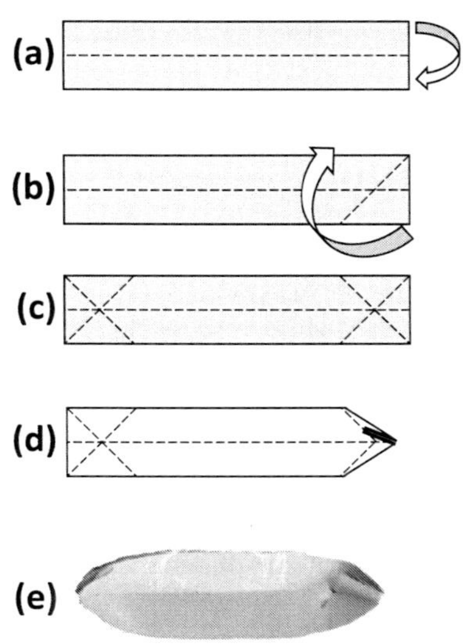

1 *Make a canoe.* Cut a strip of the paper as wide as your plastic tape and about 7–8 cm long. Place it on a surface that the tape will easily pull off (wood, ceramic, etc.), pull about 8–9 cm of tape off the roll and stick it onto the paper, matching the side edges. Peel the result off the surface and trim the two ends with scissors to remove surplus tape. (a) Fold the strip, plastic side in, in half so the long edges meet and run your fingernail along the fold to make it crisp. (b) Now, again with the plastic side facing in, fold each end diagonally from the corner up and over, first to one side then the other, lining the top edge of the end with the top edge of each side, using your fingernail to make a crisp fold line. (c) You should now have X-shaped folds at each end of your strip. (d) Get a couple of small pieces of your sticky plastic tape and tack them onto the edge of your work surface. Now fold one end of the strip, pushing

the midline fold upward and the diagonal folds at the top part of the X downward. This will produce what looks like the bow of a boat with a flap folded inside the boat. Stick one of your pieces of tape around the bow to hold it in place (use the other as well if necessary). Repeat for the other end of the strip. (e) So, you now have a canoe. Canoes are notorious for turning over, so you will need to add a bit of support. The simplest way is to put a bit of Blu Tack on the surface and push the canoe in – or add outriggers with the plastic tape. Your canoe should sit in the normal boat orientation when placed on a table.

2 *Set up the fluids.* Set up your large spoons so that they can easily hold a teaspoon of liquid. They may do this when flat on the table or you may need something like a table mat under their handles to get their mouths level enough. Put one teaspoon of water in the canoe, and one in each of the spoons. If the canoe does not have water from end to end, add a bit more until it does. Now, take your straw, stick one end into one of the dye bottles and lift it out. It should have about a drop of dye in it. If not, try again and maybe put your finger over the other end of the straw to hold the liquid in, but make sure you don't take more than about a drop (so the volume in the end of the straw looks square). Put the tip into one of the large spoons and mix until it looks uniform (and should have a much nicer, clean colour!). Repeat with the other end of the straw and the other spoon for the other colour.

3 *Do the experiment.* (a) Set up your stopwatch so all you have to do is press 'start'. Take a drop (as in 2) of dilute green and put it carefully down one end of the canoe (the idea is to get a simple blob of green into the water as far into the end as possible). Repeat at the other end with red. Start the stopwatch. Measure (as well as you can) the distance between the fronts of the two colours. (With a ~6 cm canoe, I got this as ~5 cm.) (b) Good science takes time, now you wait! You should (just) be able to see the green and red fronts slowly moving towards each other and eventually meeting. Press 'stop' on your stopwatch when they touch, check the time (I got approximately ten minutes). Wait a bit more. In not too long, a brown mixing line will form. Wait quite a while longer and it will widen (much slower than the red/green fronts moved).

4 *Do the analysis.* (a) Red and green moved similarly in opposite directions along the canoe; also, they moved up the sides of the canoe and did not stay in the bottom of its V shape. They were not driven by gravity. (b) They moved fast until they met each other and then slowed down (they are quite similar molecules) – but then they slowly mixed (as shown by the colour change to brown) so they were still diffusing

but slower than before. So, relative concentration of molecules matters for diffusion rate. (c) In my case, with 2.5 cm (for each colour) in ten minutes, I have 2.5 mm/minute. That didn't feel very fast when you were watching! How is this consistent with fast neural action?

5 *Do the theoretical integration.* (a) My 'green' dye (according to the ingredient list on the bottle) was actually a mixture of yellow (E102,[4] tartrazine, molecular weight, MW = 534.3 g/mol) and blue (E133, erioglaucine, MW = 792.9 g/mol). But given that diffusion rate depends on the inverse of MW, there is little difference in diffusion between the two – so, in this version of the experiment, you always see green as they diffuse together. The red dye (E124, Ponceau 4R, 604.5 g/mol) moves at pretty much the same rate as the green pair (average MW = 667.5 g/mol). So, for MW ≈ 640, we have 5 mm/minute. (b) What does this mean for channels? A channel is an open/closed hole in a cell wall. When open, its length has to be about the thickness of the wall. A typical cell lipid bilayer is about 8–10 nm[5] thick.[R2] So, a molecule of one of our dyes would take about 0.2 ms to get through, if diffusion were the same rate as in the canoe (2.5 mm/min = 41.7 nm/ms). In practice, it would be affected by temperature and a range of other factors. Since diffusion rate is inversely related to MW, Na+ (MW = 23) would take less time than that.

Different channels can pass different ions depending on the size of their particular hole and whether the channel has an electrical charge or not. *Voltage-gated* channels open when the voltage inside the membrane passes a critical value. *Receptor-gated* channels open when a specific shape of molecule binds to a specific shape of receptor linked to the channel (see Figure 6.1).

With a lot of sodium outside relative to inside, opening the sodium channel will allow Na+ to flow *in*, simply to balance its levels (conversely for K+, high inside, flowing *out* through a potassium channel). It's a bit like opening the flood gate on a dam and the water moving to balance its levels – except for the fact that Na+ can be high on one side when K+ is high on the other side, and so their flows are in opposite directions.

A channel is a circuit and a switch

An open door between the inside and the outside of the cell completes the battery's circuit. Importantly, unlike a conventional prison cell that only has one door, biological cells have multiple channels. They are of different sorts (think of them as having immigration guards permanently checking visas).

Those that let Na+ through don't let Cl– through, and likewise for potassium (K+), and calcium (Ca++). This specificity allows channel blockers to be selective so that you can, for example, use calcium channel blockers to reduce blood pressure (https://www.mayoclinic.org/diseases-conditions/high-blood-pressure/in-depth/calcium-channel-blockers/art-20047605), without affecting your sodium, potassium, and chloride systems.

The crucial bit here is that Na+ is an ion. When Na flows in to match Na levels, it takes its '+' charge with it but no '–' charge. So not only is there a flow of sodium (as an element) but there is a resultant flow of positive electric current into the cell. The changes in charge inside (positive up) and outside (positive down) can cause a voltage-balancing flow of negative electric current (electrons) through the liquid inside and outside the cell. The same thing happens when you connect a light bulb to a battery. (To prove current flows through salty liquid not just metal wires, try Experiment 3.2.) As you know from your landline (if you still have a copper one for phone/internet), electrons can conduct information very fast, whereas (Experiment 3.1) ions diffuse through solutions and move much slower. The brain has various tricks to take advantage of the higher speed of electricity.

Experiment 3.2 Salty liquid can carry electric current

You have probably been warned about not dropping electrical equipment in your bath (or being wet anywhere near electricity). But that is high-voltage alternating current. What about low-voltage direct current? Could that flow through your neurons? To check the capacity of the salty water inside and outside your cells to carry current, try the following [you will need a 1.5 V battery (usually AA), a 9 V battery (usually the rectangular ones with clip on terminals), a length of insulated wire, two identical straight sided water glasses, two teaspoons of salt, hot water from a kettle]:

1 Dissolve two teaspoons of salt in one 250 ml glass of hot water from the kettle (hot water will dissolve the salt faster; don't have it boiling and be careful not to burn yourself; use a glass not a mug because you want to easily see what is going on).
2 Fill the second identical glass with water from the same kettle. (The emphasis on identical/same is because in science if you don't compare the same things that have only one difference, the result can be because of some uncontrolled factor not the one you thought you were manipulating. Particularly, in this case, temperature.)
3 Get the 1.5 V battery, and the insulated wire. Cut the wire into two pieces. Remove a bit of insulation from each end of both wires. Trim one end of each so there is only a small bare bit.

4 Now hold the long bare end of each bit of wire onto one of the battery terminals (you can pinch the whole thing between finger and thumb but using two hands is likely easier). Don't let the two ends of one wire touch both terminals; you could burn yourself.
5 Dip the two short bare ends into the plain water. You may see bubbles forming with the 1.5 V battery if you wait long enough and the water is hot enough.
6 Now dip the two short ends into the salty water. You will see a stream of bubbles appear off one wire.
7 Repeat the experiment with the 9 V battery. You will see some bubbles even in the plain water and more in the salty water.

Technically, you are breaking up the water (H_2O) and salt (NaCl) into their ions (H+, Na+, O–, Cl–), and these then form the gas molecules (H_2 in both cases, O_2 with plain water, and Cl_2 with the salt solution). But the key thing is that you can see something happening that has to be caused by an interaction between the two wires through the liquid. So, the liquid is conducting. Equally important is that the salty solution does this better and the fluids in your body are around 0.9% salt.

How to flick the switch

In terms of how a light switch works, all I have described so far is how bringing electrically conducting contacts together (opening a channel) allows electricity to pass. But I have not explained either how the switch gets toggled (with our prison cell analogy, this would be how the door gets opened), nor how the electricity gets to the light, nor how it makes a light glow.

Even single-celled organisms have chemical receptors on their outer surface to detect important aspects of their environment (like food or poison). A key point that we will go into in more detail when asking how drugs affect learning is that receptors are built rather like locks. They have a specific shape that matches the external chemical they are built to detect (Figure 6.1). The chemical acts as a 'key' to operate them. This is also true of signalling chemicals released by the cell, many of which are also used by neurons as neurotransmitters. With channels in neurons (or neuroid cells, as in sponges), the magnet-like binding of a chemical to its receptor can change the shape of the receptor molecule and so pull the channel open. The resultant ionic and electrical changes can then affect the cell at low or high speed, respectively – with drugs increasing or decreasing these effects at the receptor.

An important feature of the neurons that make connections within your brain is that they can integrate (in the sense of simply adding together) the effects of the opening of very large numbers of channels on their dendrites.

How the electricity gets to the light

Neurons have two important features for signal transmission. First, a long, thin, communicating fibre (the axon), so the neuron can send a signal a long way. Second, its transmission is mostly electrical and therefore fast. (Ion diffusion through channels is slow, see Experiment 3.1.) So, the axon simply passes information from receptors that detect chemicals to vesicles that release chemicals.

The resting voltage inside the cell body is typically 'polarized' at about −60 mV relative to the outside. When lots of receptors open lots of Na+ channels, this sends positive charge down through the dendrites and brings the cell body voltage closer to zero (i.e., 'de-polarizing' the cell). When this depolarization reaches a particular threshold, it can open voltage-gated Na+ channels at the beginning of the axon (called the axon hillock).

This opening causes more depolarization, opening more voltage-gated channels further down the axon, producing an 'action potential'. Action potentials occur in many types of non-neuronal (including plant) cells and for many different purposes[6] but for us the most important is that it can send an electrical signal down the biological equivalent of a wire from your head to your toes.

A Murray Ball cartoon, reproduced with permission from Diogenes Designs Ltd. For further Footrot Flats cartoons see https://footrotflats.com/

A simple cell wall is made of two thin layers of fat, which is not an ideal insulator, and so, the electrical charge can leak out. This leakage through its insulation happens with copper undersea phone lines too, but to a much lesser extent because their insulation is so much better; nonetheless, copper phone lines need a repeater station to boost the signal about every 70 km. The neuron uses a similar trick. The long, thin communicating section (the axon) has large numbers of voltage-gated sodium channels. These channels open whenever they experience a sufficient positive change inside the cell (i.e., when the membrane potential passes threshold). The initial Na+ influx will have been due to activation of a receptor. This Na+ causes a positive change inside the cell and a high-speed electrical flow down the axon (remember Experiment 3.2). When this reaches a voltage-gated channel, this opens, lets in sodium, and generates more positive charge – and so on to the next channel, like a row of dominoes,[7] until the action potential reaches the end of the cable.

Importantly, while channels change membrane voltage by different amounts depending on the number of channels currently open, once the voltage goes past threshold at the axon hillock, the action potential will continue in an all-or-none fashion. It can be blocked by damage or by local anaesthetic. If you have had a local anaesthetic at a dentist, you will have experienced the block of sodium channels in some of your nerves. It gets all types of nerves in the area of injection so you will not only have lost pain but all other sensations from the areas connected through that nerve bundle – so be careful not to bite your lip!

With long neurons (spinal motor neurons can be a metre long), this domino system has two problems. First, it is relatively slow since it takes time to open channels and 'recharge' the action potential. Second, opening lots of channels (and then having to pump the Na+ ions back out) costs a lot of energy. The solution to both problems is to wrap a non-neuronal fatty cell[8] in a tight spiral around the axon, adding insulation (myelin). This stops electric current leaking out so much and makes the current pass down the axon further before it needs to open a voltage-gated sodium channel and get recharged.

Turning the light on

Cells can contain little balloon-like structures (vesicles) that have a fatty outside and that contain fluid and important chemicals. Some of the vesicles can fuse with the cell's outer membrane. Then, when stimulated by ionic or electrical charge, they release the chemicals they contain. As a result, in organisms that do not have nervous systems, detection by a receptor of the odour of a predator can cause one type of vesicle to release a defensive toxin, while detection of the odour of food can cause another type of vesicle to release a digestive enzyme.

With the simplest multicellular animals (see Chapter 4) cells can be specialized. Cells on the upper part of the body (that are at risk of attack) can release toxins and cells on the lower part (which is where the stomach would be if they had one) can release digestive enzymes. These animals are small enough and flat enough that they don't need nerve-like cells to coordinate their actions.

With neurons, an action potential travels until it reaches the end of the axon, where the vesicles are, and causes them to release their contents – in the simplest cases, into the outside world or onto a muscle. With neural nets, the vesicles often release a neurotransmitter chemical to affect the next neuron in line via receptors in its dendritic tree.

Neurons – the big picture

Sensory neurons have specialized detection mechanisms rather than dendritic trees but are otherwise like other neurons (including motor neurons), which follow a fairly uniform plan. This (plus some microscopic details) is shown in Figure 3.2.

At the start (Figure 3.2A), a transmitter chemical (shown by a T shape) binds to a matching receptor (shown by a shadow T shape because it's the matching of the shapes that determines the binding). This causes the receptor to deform, pulling the ion channel open (double-headed arrow in the figure). Sodium (Na+) rushes in – depolarizing the neuron (making its potential less negative than when the channel was closed).

Normally there are lots of T molecules opening lots of channels. The positive charges cause current to flow in the dendritic tree (Figure 3.2B). The currents sum at the cell body (Figure 3.2C), which effectively calculates how much T arrived throughout the dendritic tree. That is, it works out the strength of the input signal.

If the cell body depolarization is large enough, it goes past the triggering voltage required to open voltage-gated Na+ channels at the axon hillock. This starts an action potential. In a myelinated neuron (Figure 3.2D), current flows along the myelinated section and restarts the action potential at the unmyelinated gaps.[9] Once started, the action potential travels to the end of the axon (which can be up to a metre long; Figure 3.2E).

At the end of the axon, there is a lot of terminal branching (Figure 3.2F) to allow multiple synaptic contacts with the dendritic trees of a large number of target neurons. Only one target is shown in the diagram, but there are usually a lot – and two collaterals both with branches are shown in Figure 3.2G to help you keep this in mind. The depolarization causes release of neurotransmitter, which then binds to receptors in the dendritic tree of the target neuron (so we are back to the equivalent of Figure 3.2A).

The transmitter is held in vesicles inside the synaptic 'bouton' (shown as a flattened blob in Figure 3.2H). Many of these vesicles are located at the membrane on one side of the synaptic gap. The depolarization causes them to burst open, releasing T into the gap. This diffuses and, if it binds to a receptor, causes a sodium (or other) channel to open. It only stays bound for a short time and then carries on diffusing. It ends up getting sucked back up into the terminal and repackaged in a vesicle (example at top of synapse) or escaping out of the synaptic cleft and being broken down by special chemicals (enzymes[10]). This is the basis for the effects of drugs that block reuptake or breakdown.

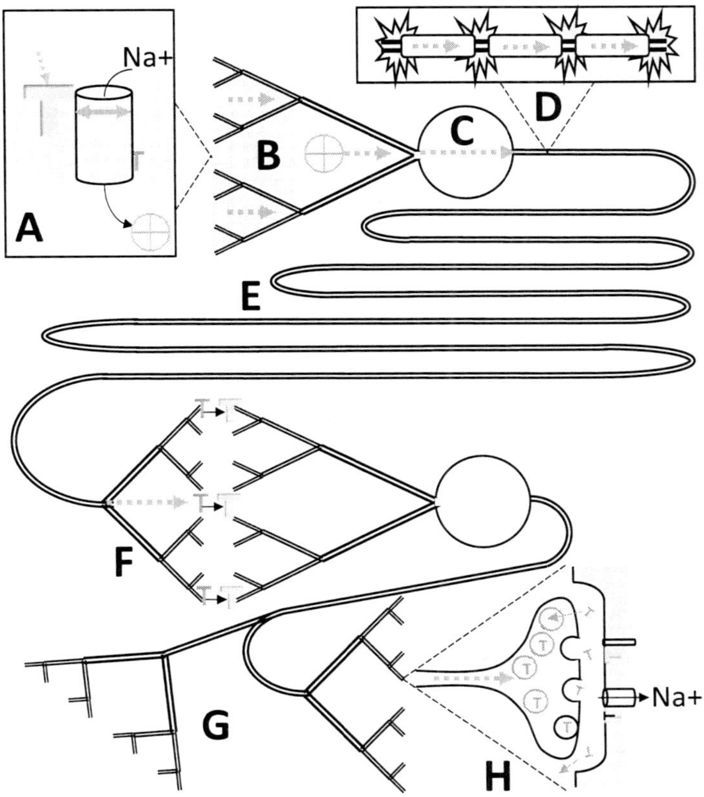

Figure 3.2 The neuron – the big picture. See the text for a description of how this operates. (A) A very greatly magnified picture of a receptor site and channel. T represents a transmitter molecule. T represents its receptor site (for more on receptors and their types, see Figure 6.1). ⊕ represents the positive charge produced by the inflow of sodium (Na+). (B) Dendritic tree gathering charge that flows (→) to (C) the cell body. (D) Magnified picture (showing myelin insulation and ✸, regeneration of action potentials) of part of (E) the axon. (F) Terminal of the axon showing arborization to connect with dendritic tree of next neuron. (G) Two collaterals from one neuron. (H) Greatly magnified picture of a synaptic bouton with ● vesicles.

Here we have looked at the most basic mental processing units. But, for mental function, their connections and organization into functional units are equally important. We cover these in the next chapter.

Notes

1 Glia also play a role. They influence communication and plasticity at the synaptic and network levels. They are not just there to support neurons.[R3]

2 The sponge can be viewed as a colony of distinct cells, you can put it through a sieve and it will put itself back together (https://en.wikipedia.org/wiki/Sponge), but to behave the way it does these cells need to communicate and they do so as we will see basically the way neurons communicate.

3 An ion involves a single element in solution and has an electric charge indicated by +, –, ++, etc. In solid table salt Na and Cl are a molecule NaCl, where their + and – lock each other together and balance out their electric charge. But, in the liquid outside and inside a cell, their ions (Na+, Cl–) can move independently.

4 I looked up what E102 was on Google and then, separately, looked up the MW of tartrazine. The food companies don't tell you too much!

5 1 nm = 10^{-9} m (1 billionth in US terms, 1,000 millionth in European terms).

6 Shepherd[R4, pp. 100–1] lists fertilization, cell division, bioluminescence, motility (ciliary, vascular, and muscle) as well as a wide range of neural and hormonal effects.

7 https://www.youtube.com/watch?v=FWgH0hXZKrE; or search 'falling dominoes' for lots of examples.

8 The brain has lots of non-neuronal cells, called glia, that hold the brain together. 'Glia' come from the ancient Greek word for glue, γλία. The cells that wrap round neurons are glia called Schwann cells.

9 These gaps are called Nodes of Ranvier and are much like unmyelinated nerves. It's the insulating myelin that is really important for conduction and which is where the problems come from in demyelinating diseases. Google lists over 20 of these, including multiple sclerosis, tick-borne encephalitis, optic neuritis, and Guillain–Barré syndrome.

10 These enzymes are mostly named according to what they break down plus and 'ase' ending. So, monoamine oxidase breaks down monoamines (noradrenaline, dopamine, serotonin), acetylcholinesterase breaks down acetylcholine, and dopamine-β-hydroxylase (DBH) hydroxylates dopamine at the beta position converting it to noradrenaline. DBH deficiency can cause a range of blood pressure and mental problems https://en.wikipedia.org/wiki/Dopamine_beta-hydroxylase.

References

R1. Herculano-Houzel, S, The human brain in numbers: A linearly scaled-up primate brain. *Frontiers in Human Neuroscience*, 2009. **3**: A031.

R2. Regan, D, Williams, J, Borri, P, & Langbein, W, Lipid bilayer thickness measured by quantitative DIC reveals phase transitions and effects of substrate hydrophilicity. *Langmuir*, 2019. **35**: 13805–14.

R3. Allen, NJ & Lyons, DA, Glia as architects of central nervous system formation and function. *Science*, 2018. **362**: 181–5.

R4. Shepherd, GM, *Neurobiology*. 1994, Oxford: Oxford University Press.

4 Building the machine

How did the mind machine happen?

To apply mental mechanics, you need to know how the micro-processing units work (Chapter 3) and how they are organized into circuits at the macro (structures, architecture) level. Both micro and macro expansion during both evolution and your development were constrained. This is the key to how your mind is built. Unlike the parts of a computer before they were assembled, a growing brain must work at all sizes. How can this happen? Why did cells become neurons? Why did clusters of neurons become brains?

To expand a mind machine, take the current version and paste over a new bit – like icing on a cake. Neurons are the result of simple additions to non-neuronal cells. Brain nuclei are the result of clustering neurons. Brains are the result of clustering nuclei – and then adding more layers. What does this look like?

Take a deep breath. We are about to go up several orders of spatial magnitude – from molecule micro to muscle macro. We are also going to start before there were neurons. Meet the flat blob animals.

From 'flat blob animals' to sponges to molluscs

Placozoa ('flat blob animals'[1]) "are the simplest animals on Earth". Nonetheless, they graze on algae like cows on grass and may even have a sleep-wake cycle[2] like us.[R19, p. 528] Amazingly, "without any recognized neurons or muscle, placozoans exhibit fast effector reactions and complex behaviors. ... [Their] rapid action potentials ... are sodium-dependent ... [with] 5–7 different types of voltage-gated sodium channels".[R20, p. 120]

So, despite being not much other than flat and blobby, they can keep food-filled water moving around them with small whip-like organs. They can also move around. They then feed by clamping themselves down and releasing enzymes from their underside (basically a half-stomach with ground the other half) to digest what they are on top of. Then they suck it up. They can also release toxins from their upper side as a defence.

DOI: 10.4324/9781003465027-4

All this is controlled by action potentials[3] and the release of signalling chemicals, e.g., glutamate (as in MSG, the Chinese wèijīng and the Japanese aji-no-moto), glycine, GABA,[4] and ATP.[5] These signalling chemicals have extensive variation within and across nerve-free animals. They appear to have been co-opted later in evolution as classical neurotransmitters.[R18]

Porifera (sponges[6]) are a neat model of the next step up. They not only have whip-like organs in their centre tube to move water (as placozoans do on their outside) but also a layer of cells that can contract (like muscles, to move more water) and that make signalling contacts among themselves (unlike our muscle cells but somewhat like our neurons). They also receive contacts from cells on the outside surface of the sponge. These multicellular interconnections can almost be viewed as a nervous system.[R17] But the cells are neuroid[R15] rather than being true neurons – the key middle-layer cells *both* communicate *and* contract.

Cell–cell contact to improve movement control beyond simple chemical release (which, nonetheless, humans still have via hormones) appears to have evolved at least five times. But only two of the lineages progressed from neuroid cells to nerve nets and then central nervous systems[R15]; and the sponge lineage (Porifera) is distinct. In a sense, the sponges have a communication system that is *too* efficiently matched to their sedentary 'grazing' lifestyle to evolve to a neural one. Neurons are expensive and would give them no additional advantage. That is why thousands of species of sponge are still here and thriving but do not have brains, nor seem to miss them. If placozoa have zero neuronal layers, we could say that sponges have half a layer (i.e., one layer but of cells that are only halfway to neurons). In contrast, the more predatory comb jelly (Ctenophora) lineage evolved neurons. These can have 1, 2, 3 or more layers of neurons (Figure 4.1). At each step the extra catches were worth the extra effort.[7]

The sea anemone is an example of a one-layer-neuron system. Some species of jellyfish are examples of a two-layer-neuron system. Other species of jellyfish and molluscs are examples of three-layer-neuron systems.[8]

So, to expand to much larger systems like our own, simply increase the depth and number of intermediate layers. Note that the zero neuron, flat blob already has a mass of detectors, effectors, and chemicals at its disposal. You add neurons, in ever-increasing layers, to increase the complexity and sophistication of the 'data processing' that culminates in each 'decision' to act.

Wetware versus hardware/software

All this 'wetware' is expensive to run. That is why sponges don't bother with it; and why fish that live in permanent darkness in caves have lost their vision. Equally importantly, it has to fit in with biological design rules not those typical of electronic computers. This leads to a number of important differences between a brain and a computer.

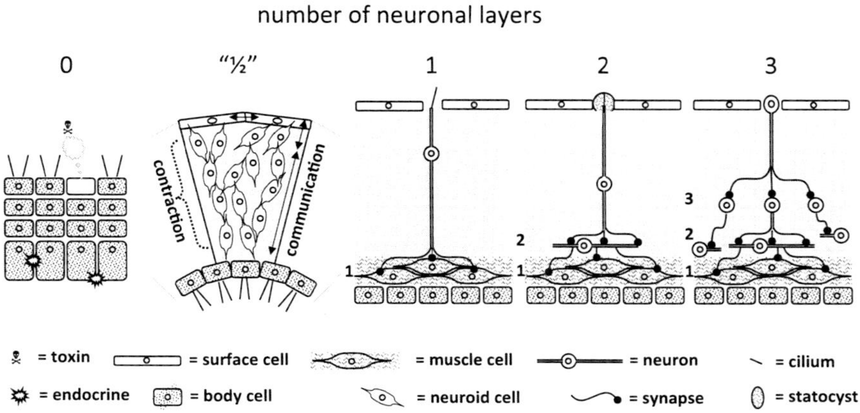

Figure 4.1 From 0 to 3 neural layers. (0) ('flat blob animals'). As described in the text, with no neurons or muscles, these can still move about, release toxins to defend themselves, release enzymes from their lower surface to digest their food, and use cilia on their upper surface to move water (and food). **'1/2' (sponges)**. These have neuroid cells that can contract (like muscles) and communicate via cell-cell connections (like neurons). They can move water and food like flat blobs, but this is on their inside tube not the upper surface. (1) **(sea anemone)**. These have a single neuron (often with a cilium to detect movement) that fires, releases neurotransmitter from synapses, and so contracts muscle. (2) **(some jellyfish)**. These have true sensory neurons. I have shown one with a statocyst. This is a hollow sac that can statolith particles in it. Its movement fires the sensory neuron, which then fires a motor neuron, which releases transmitter to make the muscle contract. (3) **(other jellyfish and molluscs)**. These now have intermediate neurons between the sensory and motor ones, allowing much more complex integration. **Lots (not shown)**. You can see it's easy to build every more complex systems by adding layers. But the changes are never at random. Steady addition of neurons must deliver an advantage at every step and not disrupt the system.

The first important difference between neural systems and computer networks is that during both evolution and development (which follow similar paths) the 'machine' has to function perfectly at all stages of expansion. So, the development of layers involves smoothly graded expansion and differentiation of cells into different types.

The second important difference from a computer is that the wetware processes and stores information by its position in the brain not by a digital code. In contrast, a computer can store the same digital information anywhere in its memory. Because of this position coding, essentially the same pattern of neural activity will produce perception of a burst of light in visual cortex, a blast of sound in auditory cortex, etc. So, the same electrical stimulus delivered to different parts of a patient's brain will produce different sensations – which allows surgeons to have a good idea of where an electrode has reached when they are lowering it into the brain. (Don't try that yourself at home.) Because of this progressive development and positional coding, the brain's connections are topographically organized. That is, adjacent points in one structure are connected to adjacent points in the next. In the simplest cases, you can think of this as a simple point-to-point mapping.

A third important difference is that, because of both graded change and topographic organization, brains have to grow, and to be constructed in a specific, hierarchical way. Basically, when evolution or development need additional processing power, they just steadily increase the size of the top layer of a structure until a new structure can split off the top of what is there already. This results in hierarchical organization of systems with the lowest levels retaining much of the capacity for original, primordial functions and the higher levels working by controlling the lower levels.

How neural layers develop

So, how does 'adding layers' happen smoothly rather than in a discrete, 1-2-3, fashion? Also, how can it involve the changing of cells from one type to another? These changes occur both during evolution and development. To get the picture, look at the human cortical development shown in Figure 4.2.

First, look at the cell plot through the middle of the figure. At specific times during development, we can see one, two, or more layers matching the layers we illustrated for evolution in Figure 4.1. But in between, you can see that there is a steady transition (which will also have occurred during evolution). During this transition, you not only get increases in the number of cells (symmetric division) but also other changes (such as transformation) that occur in cells that are forming new layers.

Next, look at the brain representations at the top of the figure. The gross brain morphology also changes – with folding and wrinkling. At the third brain along, the human cortex is smooth, like an adult rat; it only wrinkles later (at the fourth one), like an adult chimpanzee. Anatomy pictures, particularly of primate cortex, can look very complicated because what would be a nice smooth surface in a small rat brain has to be folded up (like a shirt being stuffed into a crash helmet to get both into your suitcase) to fit inside the skull (which is the brain's crash helmet and needs to be helmet-shaped for structural reasons). Later, I will use diagrams that have the surface of the cortex de-crumpled. This will make the functional organization clearer. That said, the convolutions are not just a matter of fitting into the skull, they also help with various aspects of network connection.

Finally, let's look at the bottom row of images. So far, layer development has probably all sounded very theoretical. But, at each stage of this development, one can get problems. Microencephaly (micro brain), hemimegancephaly (expanded half brain), polymicrogyria (too many small folds), and a range of more complicated others.[R24] Yes, these are 'neurological' disorders

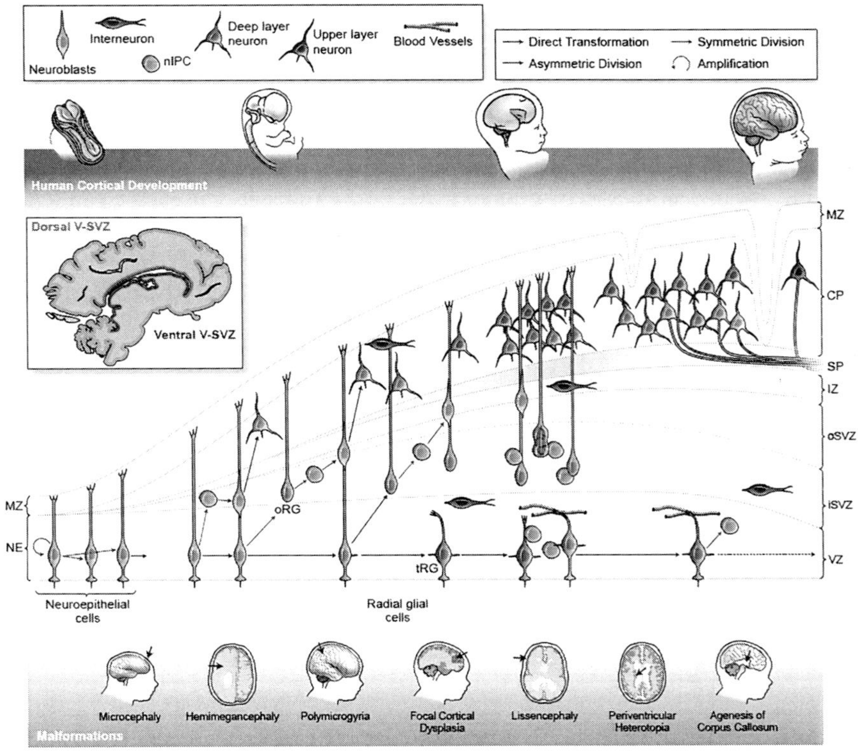

Figure 4.2 Human cortical development and stages of malformation. For details, see text. Figure from Subramanian et al.[R24, Figure 1, p. 3] reproduced under the terms of the Creative Commons Attribution License (CC BY).

but, nonetheless, they change mental function often drastically. Subtler variations, coupled with experience, make each of us have a different brain and personality.

Wiring a brain is not like wiring a computer

If you think of the cortex as like icing on the top of a cake, we just looked at how the depth of the icing can change. But, what about the organization on the surface of the cake? Critically, the brain codes information in terms of position.

For example, the retina of a vertebrate's eye can be thought of as a simple hemispherical flat sheet of light receptors (and associated local circuits) that is much like a digital camera. Of optical necessity, adjacent positions on the retina map to adjacent 2D positions in the observed environment.

During development, the retina sends fibres to connect to the tectum (which does the main initial visual processing) and these fibres find their way to a spot that 'matches' where they came from (Figure 4.3A). Note that this 'upload cable' sends all of the information about an image at once, in parallel, not like the serial stream of information that transfers an image from your camera to your computer down a USB[9] cable.

The wiring-up between the two 'sheets' of cells is done using very simple rules based on chemical gradients.[R12,22] One chemical can code from top to bottom (shown as shades of grey in Figure 4.3) another can code from left to right (shown as different circle sizes in Figure 4.3). In each case, fibres can be thought of as lining themselves up like a rank of soldiers asked to stand to the left of anyone taller than they are. The soldiers don't have to be told to stand in an exact place – they will find the right place using the rule. So, in terms of how the system works you can almost always think of the brain as wired basically as in Figure 4.3A – but with the complication that the target (in this

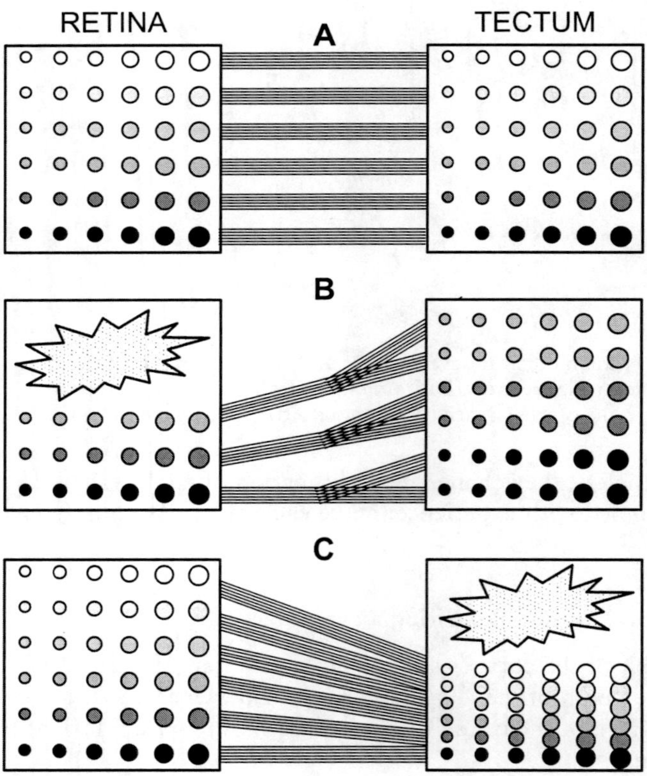

Figure 4.3 How to wire a brain. (A) The connections of the retina to the tectum are organized in terms of chemical gradients coding two dimension (darkness of shading versus circle size).[R12,14] (B) The gradients can sometimes still work so all the tectum is still wired up even if half the retina is damaged. (C) Likewise all the retina can wire up even if half the tectum is missing.[R4]

case the tectum) may be bent rather than flat (the crash-helmet problem) and be rotated in 3D space relative to the fibres' origins.

In frogs and goldfish, the way this wiring works can be studied via re-wiring when the optic tract is crushed and then the retina or tectum is damaged. Depending on species and youth, one can observe a half retina re-wiring into a whole tectum (Figure 4.3B) or a whole retina re-wiring into a half tectum (Figure 4.3C).[R4] However, in older animals or other species (particularly us), there is less plasticity and the remaining connections follow the pattern of Figure 4.3A. The apparently more flexible form of re-wiring may, in fact, depend on the change of the chemical gradients during regeneration rather than a change in the chemical target for which a specific fibre is aiming.[R14]

An advantage of this simple method of wiring itself up is that the brain is relatively resistant to incidental damage during development. For example, if you lesion half the retina before the wiring-up has taken place, the remaining half retina can sometimes still wire up the whole tectum (Figure 4.3B). Conversely, if you lesion half the tectum, the whole retina can cram all its wiring into the remaining half (Figure 4.3C) – so while the resultant vision might not be the highest quality, the frog or goldfish will still have a full field of view and a good chance of catching its lunch.

With the simple 2D mapping of the visual world into the retina, it makes intuitive sense that the retina should be 2D mapped into the tectum, and easy to understand how this could be controlled by two chemical gradients, one for each spatial dimension.

It's not so obvious that the rest of your brain would be built this way, too. However, with steady evolutionary (or developmental) expansion, it makes sense to keep the same basic chemical methods operating, adding extra signalling molecules for extra structures. At least in its early development, your cortex was patterned by signals coming from distinct centres, as was the thalamus, which provides the main sensory input to the cortex.

The resultant 'primordial' maps will then have been shaped by interaction between the cortex and thalamus[R6] and by the impact on both of early developmental experience, which we will discuss a bit more at the end of Chapter 6.

Brain systems expand as hierarchies

Topographic mapping, plus the fact that the systems must function while they are expanding, drives an important bit of mental mechanics. Brain systems are organized hierarchically.

To see how this comes about look at Figure 4.4. Initially, a brain system starts as a single lump (a) that is more a processing unit than a system. It will contain interconnected neurons that carry out calculations but these can all be seen as being inside one 'nucleus' and doing much the same job. Evolutionary pressure can push this nucleus to expand, allowing some cells to start

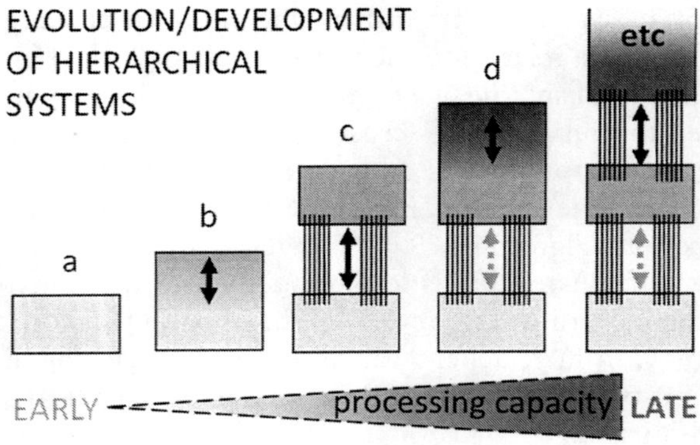

Figure 4.4 Expansion into hierarchical systems. This shows how one can go from a simple primordial clump of neurons (A) via expansion and differentiation (B), see also Figure 4.2 to distinct connected clumps (C) to ever higher levels of control (D) and so hierarchically connected systems of nuclei. The same basic processes occur (and in much the same way, but not true recapitulation) in both evolution and development. In both cases, the progression from early to late systems involves an increase in processing capacity. See Figure 4.2 for an illustration of how cell types can change during expansion.

doing something a bit different from, and perhaps controlling, the others (shading in b).

Eventually, the original nucleus can split into two with a primordial part still carrying out simple functions much as before; and a new part carrying out additional functions and interacting with the primordial part via axonal connections (vertical lines in c).

The process can then repeat (d, etc.) with each higher level developing more cognitively complex functions by processing already pre-processed information via its interactions with the lower levels. Meanwhile (unless over-ridden by a higher level) the lower levels can carry on producing the original quick-and-dirty solutions that in extreme cases will allow survival when slow-and-sophisticated calculations would not.

Building on top: from lancelet to rat to human

For obvious crash-helmet reasons, you cannot keep adding chunks on the front of a brain. So, as you can see in Figure 4.5,[10] smaller brains have a subcortical core (created by expanding at the front) and then larger brains have cortex expanding sideways and backwards and then becoming folded. You can think of the cortex as a layer of icing (and marzipan) over the subcortical cake. As we will see, this cortical sheet is still hierarchically organized across its surface. But it also involves evolutionarily and developmentally graded expansion of its depth.[11] As we will discuss shortly, the underlying subcortex in humans retains much of the organization of the brains of animals that have little or no cortex.

The top 'brain' in Figure 4.5 is that of a lancelet (amphioxus). Lancelets are very small and fish-like. But they are invertebrates (no back bone) and their organs are more primitive than fish. They have a tube-like equivalent of a spinal cord (the notochord) but with cartilage rather than bone supporting it. This neural tube is somewhat expanded at the head end and, despite lacking any obvious division into segments, has zones that in terms of gene

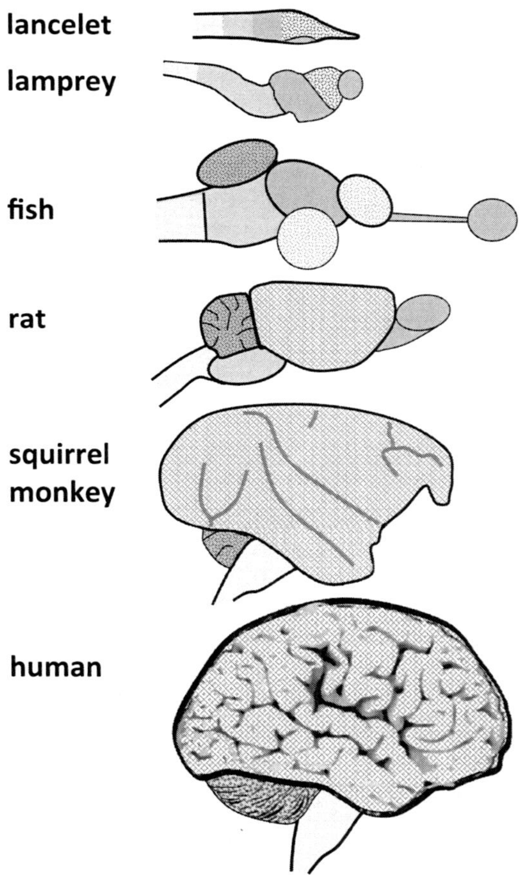

Figure 4.5 Expansion of the brain across species[10]. The lancelet (amphioxus) shows a modest enlargement at the front of the notochord (its equivalent of a spinal cord). It has zones (shading) that, in terms of genetic regulation appear precursors of distinct zones in more complex brains. The lamprey shows expansion, particularly at the front. Additional lobes are present in fish. Cortex expands smoothly over and round these in the rat. With squirrel monkey and then human, the cortex increases in area, requiring ever more extensive folding (not to scale).

regulation appear equivalent to the hindbrain (mid-grey), midbrain (dark grey), and the diencephalic[12] segments of the forebrain (stippled) in vertebrates.[R16,26]

Next is the lamprey, an eel-like jawless fish. Their larvae are quite like adult lancelets. They do not have a proper spine but, unlike lancelets, have chunks of protective cartilage above their notochord. As you can see in the figure, they have a distinct brain with fairly clear segments and, at the front, the forerunners of the cerebral hemispheres and hippocampus.[R9] The olfactory bulb (dark grey oval) is at the very front.

Fish maintain much the same basic plan but with the addition of a number of 'side' lobes: cerebellum[13] (dark grey, stippled); twin optic lobes (dark grey); and twin cerebrum and inferior lobes (light grey, stippled). The olfactory bulb (dark grey oval) is connected via the olfactory tract.

Rats have a true telencephalic[14] forebrain, with a smooth sheet of neocortex spreading up and out and then back from the front midline to cover over the entire diencephalon. Other than some expansion in size, the diencephalon, cerebellum, and olfactory bulb are very similar to a fish brain.

With monkeys (then apes) then humans the neocortex continues to expand relative to subcortex, becoming progressively more convoluted. This expansion is relatively greatest for frontal cortex and for cerebellum (which is strongly connected to frontal cortex)[R1]; and is relatively modest for the hippocampus (about 40 times less expansion than neocortex across primates),[R23] The hippocampus gets buried deep inside the temporal lobe of the neocortex. Human brain expansion during development follows a generally similar pattern. But it is only superficially the case that 'ontogeny follows phylogeny'.

Your 'triune' brain

One way to look at the human brain, then, is to see it as consisting of a *reactive* reptilian core, overlayed with an *emotional* paleo-mammalian layer (a 'visceral brain', often called the limbic system and including the primitive three-layered hippocampal 'archicortex'), overlayed again with a *predictive/cognitive* neo-mammalian layer (including the more recently developed six-layered neocortex).

Paul MacLean first presented this view in detail[R10] and then named it as a theory of the 'triune' brain:

> Markedly different in their structure and chemistry and, in an evolutionary sense eons apart, the three assemblies constitute an amalgamation of three brains in one, a triune brain. Although extensively interconnected, there is evidence that the three assemblies can operate somewhat independently.[R11, p. 126]

Thinking of the brain like this can make a range of things easy to understand. But, like thinking of evolution in terms of purpose, it's only a convenient shortcut and ignores the fact that the *relatively* conserved lower brain levels have nonetheless adjusted as higher levels developed. Note, for example, that all the visual functions of the frog are carried out by its tectum (Figure 4.3); whereas, only some of ours are in the superior colliculus and others migrated to the visual cortex. Also note that, in hierarchical systems, the higher systems operate by controlling the lower ones rather than replacing them.

So, you can look at the lower levels as a reptilian brain (capable of doing many of the things that a reptile can). But it is better to think of them as having evolutionarily conserved 'quick-and-dirty'[R7] survival circuits.[R8] When more time is available, slow-and-sophisticated limbic and neocortical circuits can take over control of the survival circuits and produce better long-term solutions. Problems like panic can occur when higher levels do not have time to inhibit lower ones.

If you try and take apart a rat or human brain (Figure 4.6A, see-through version in panel B) you will notice that you can essentially peel off the two cortical hemispheres (telencephalon) if you start at the back (Figure 4.6D). But that they are firmly joined to the (undivided) diencephalon at the front (Figure 4.6E). If you detach the front part (Figure 4.6F) and the cerebellum (Figure 4.6C), you are left with a brainstem and limbic cortex (Figure 4.6G), which, cut in half and viewed from the inside (Figure 4.6H), looks very like a lamprey or fish brain (Figure 4.4).

This arrangement is one of the reasons MacLean saw the 'reptilian brain' as distinct. It had just involved forward expansion (lancelet-lamprey-fish) as shown in Figure 4.4, with more complex processing added at the front. Fish then added bilateral sensory and motor processing units at the front. You can think of these units as then continuing to expand in a forward direction – except that in practice they have folded back over the diencephalon (and so keep the skull more like a crash helmet than a tea tray). Also, while their expansion has been at the 'front', this is where the sensory information arrives, and so functionally the newer 'isocortex' involves very simple processing of information that it then feeds to progressively older bits of cortex. All this information, then, essentially converges in its *highest-processed* form in the *oldest* 'allocortex' at the junction of the telencephalon with the diencephalon. This makes sense in terms of changing input to, but not messing with, central control.

On this view, the highest levels of mental processing have stayed in much the same location since fish – and the main difference between fish and us is that we have a whopping amount of extra sensory filtering and extra motor programming attached. This made processing in the highest levels ever more sophisticated but without the danger of directly changing the integration of multiple streams of information by those crucial higher levels.

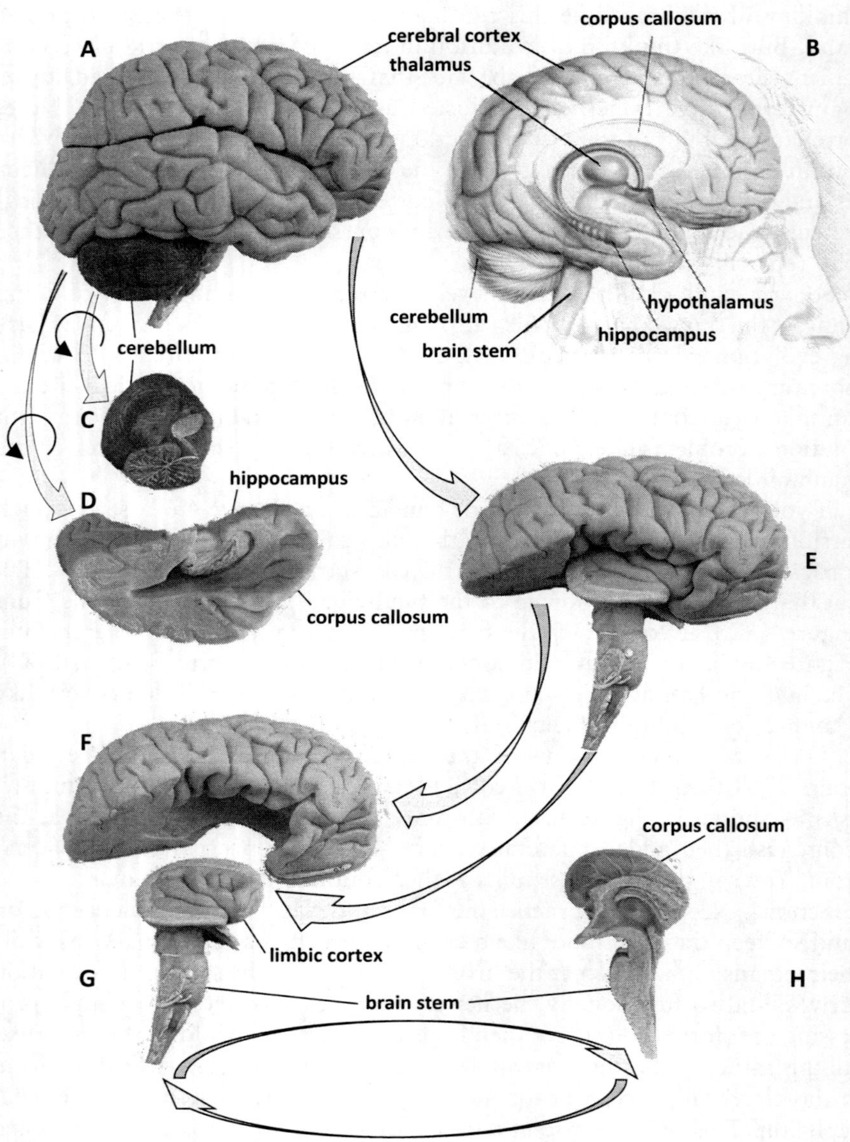

Figure 4.6 Pulling the brain apart. For details, see text. See also Experiment 4.1. Panel B is in the Public Domain (with other goodies) at https://www.wpclipart.com/medical/anatomy/brain/human_brain_drawing.jpg.html.

Experiment 4.1 Brain dissection

Figure 4.6 was made by photographing a plastic model with a cell phone camera and editing with Paintshop Pro 7. If you want to get one as shown, Google 'human brain model'. These can cost as little as $80 – make sure yours has removable parts. These 3D models are available because it really does help your understanding to take them apart, look at the bits from different directions, and put them back together again to get a good idea of how the brain is put together. Some come coloured and with detailed identification of the parts.

If you want to try genuine dissection, you will get a good idea of how the human brain is put together if you look at a sheep's brain. These should be available from a local butcher or a specialist meat supplier (e.g., https://www.continentalmeats.co.nz/product/lamb-brain/) at about the same price per kilo as other lamb cuts. Fresh brain can be very fragile so academic classes usually use preserved sheep's brains (e.g., https://www.carolina.com/preserved-sheep-organs/carolina%27s-perfect-solution-preserved-sheep-brains/FAM_228703pr?utm_source=internal&utm_medium=link&utm_campaign=featuredRecMDEX_products_404). These hold together much better and can be kept for longer. But they need care when handling because of the toxic nature of the solutions used for preservation of the tissue. For detailed instructions on how to proceed with a dissection, Google 'sheep brain dissection guide' and check sites such as https://www.biologycorner.com/anatomy/sheepbrain/sheep_brain_dissection_guide.html.

The prefrontal planning hierarchy

So far, our picture of the brain has been fairly convoluted and, indeed, crumpled. This matches the complex way people often talk about brain processing in the frontal cortex. They distinguish: motor control, gaze control, working memory, language, goal control, planning, etc. However, opening the cortical sheet out flat, and then looking at its organization, can lead to a simpler view of the mental mechanics.

If you look at the flat map in Figure 4.7, you should find its 2D organization quite comforting. The bulk of the apparent complications of the brain shown in Figure 4.7A are the result of cramming what is really a flat sheet (and looks like one in the case of the much smaller rat cortex, Figure 4.5) into a spherical skull.

The graded shading shows one dimension of the sheet: evolutionary/functional depth. At the far right is the oldest, structurally simplest, cortex (allocortex). As we progress to the left there are progressively newer layers of

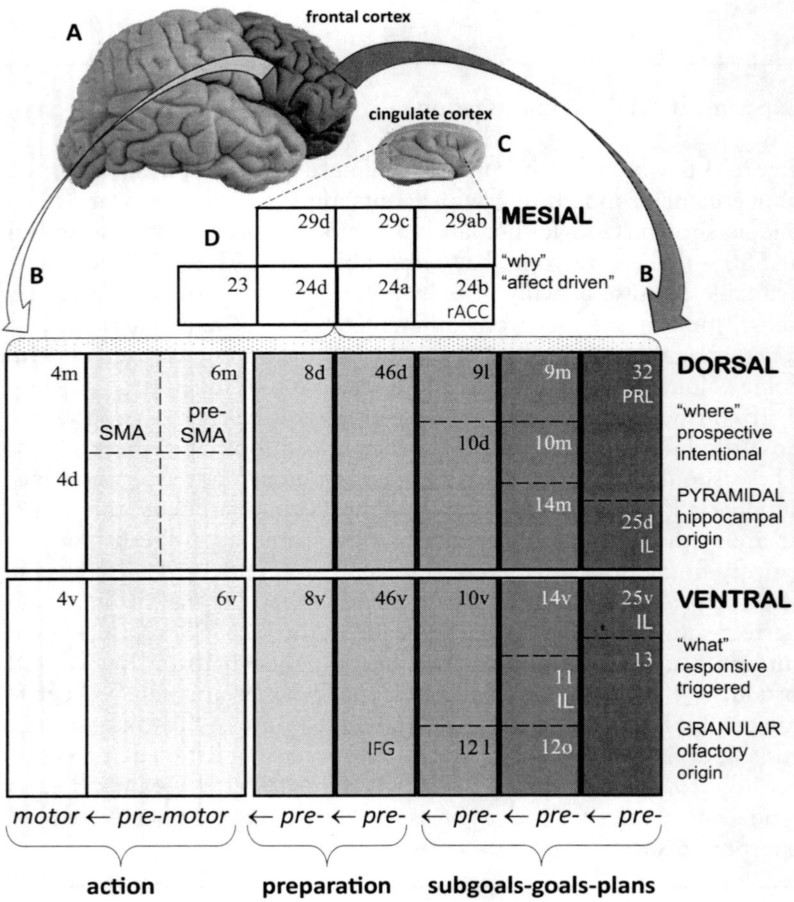

Figure 4.7 The frontal flat map. The frontal cortex (darker grey in A) can be lifted up, opened out, and unfolded (B) to produce a flat map. The version shown is a kind of average across rats, monkeys, and people. For a detailed flat map of the whole rat brain and a simple flat map of the human brain, see Swanson[R25]. Underneath (and at the midline) is the cingulate cortex (C), which can also be unfolded (D). The result is a sheet that runs from mesial up to dorsal cortex and then down to ventral cortex (including the orbital areas close to the eyes). The numbers in the flat map are identifiers of distinct cellular Broca Areas (BA) identified by Paul Broca after whom the right frontal language area (in the inferior frontal gyrus marked as IFG in the flat map) is named. The most modern cells (isocortex) are in BA4 (motor cortex that controls immediate actions), as we progress to BA6 (pre-motor cortex, including the supplementary motor area/SMA and pre-supplementary motor area/pre-SMA that prepare for action), then BA8 (frontal eye fields that control eye movement prior to targeting action), then BA46 (working memory), etc. through various levels of goal processing, the cells become more archaic (allocortex) that finish up in infralimbic/IL and prelimbic/PRL cortex. One way to think about all these areas is as a stack of modules, each of which is preparatory to the one before (for more detail, see Chapter 9) – with evolution having just added extra steps at the front to get more sophisticated motor control. Thus, the highest level of goal processing can be thought of as just pre-pre-pre-pre-pre-pre-motor control (Figure 9.7). To reach for a particular point is space, you need to remember where to look (working memory) and then look (frontal eye fields) and then work out where the target actually is (pre-motor) and then move your hand (motor) – for circuit-level details, see Goldman-Rakic[R5].

more structurally complex, cortex (isocortex). Thus, control of immediate action appears to have been continually expanded with the result that what is computed by older integrating areas receives more sophisticated and more anticipatory input.

On this view, planning is the highest level of goal processing but can be thought of as just pre-pre-pre-pre-pre-pre-motor control. Planning anticipates upcoming goals, goals anticipate subgoals, subgoals anticipate active/working memories, working memory[15] anticipates things like an upcoming eye movement, with the eye movement essentially anticipating the target selection that then precedes direct action.[16] This would place 'you', as the highest level of anticipatory integration, somewhere at the junction between subcortex and allocortex.

The other dimension of the sheet has three parallel streams: mesial, dorsal, and ventral. These separately process the why, where, and what aspects of an object: its affective value, location, and physical nature. This separation is not obvious if your brain is functioning normally. However, damage in the relevant areas can have quite specific effects with surprising results such as "The Man Who Mistook His Wife for a Hat".[R21] 'What' and 'where' seem simple enough, although their processing has some unexpected features; but 'Why' is about affect/emotion. Emotion is one of the most elusive concepts in psychology and will need the whole of the next chapter to nail it down.

Notes

1 For example, look at https://en.wikipedia.org/wiki/File:Trichoplax_adhaerens_photograph.png.
2 Evolutionarily speaking, sleep must be good! See also Chapter 8.
3 Chapter 3, *How the electricity gets to the light.*
4 Gamma-amino butyric acid, but you don't need to remember that as everyone uses GABA to refer to it.
5 Adenosine tri-phosphate, but you don't need to remember that as everyone uses ATP to refer to it.
6 For example, see https://www.montereybayaquarium.org/animals/animals-a-to-z/orange-puffball-sponge.
7 Here and elsewhere, I have allowed myself to use teleological, that is, goal-oriented, language for evolutionary changes as its simpler to understand. But it should have been phrased in terms of teleonomy, that is, a history of chance and necessity that just happened and was in no way planned.
8 Nauta et al.[R17] cites George Parker's (1919) *The Elementary Nervous System* for details of 1, 2, and more simple neuron nervous systems. This runs to over 200 pages and has been reproduced recently as a classic text. For a facsimile copy, see https://www.biodiversitylibrary.org/item/17085#page/9/mode/1up. For the reproduced copy, see https://www.amazon.com/Elementary-Nervous-System-George-Howard/dp/1103394738.
9 USB = Universal *Serial* Bus. That is bits of information are sent one after the other. Some computers also have *parallel* ports but compared to the brain these have very small numbers of parallel channels.
10 The mammal diagrams are based on a collection of photos at http://brainmuseum.org/index.html showing a wide variety of mammals' brains (e.g., platypus) from a range of perspectives. These photos are from the University of Wisconsin and Michigan State Comparative Mammalian Brain Collections, as well as from those at the National Museum of Health and Medicine. Preparation of all these images

and specimens was funded by the National Science Foundation, as well as by the National Institutes of Health. Amphioxus is based on[R16] and the lamprey on[R9].

11 In the adult brain the combination of the two processes can result in 'transition zones'. A particularly clear example is where the archicortical (old cortex) hippocampal layers expand, via the subiculum (https://en.wikipedia.org/wiki/Subiculum#/media/File:CA1_to_subiculum4.jpg), to the neocortical (new cortex) layers of the retrosplenial cortex.

12 Diencephalon (between brain) is the part of the forebrain that connects the brainstem (midbrain and hindbrain) to the telencephalon (far brain), which is the front part of the forebrain and predominantly consists of the cerebral cortex.

13 The cerebellum (little brain) is so called because it is largely separate from the rest of the brain and is the earliest part to have cortical layers and to get convoluted. It looks rather like a cauliflower with only the main stem connected to the hindbrain.

14 Diencephalon (between brain) is the part of the forebrain that connects the brainstem (midbrain and hindbrain) to the telencephalon (far brain), which is the front part of the forebrain and predominantly consists of the cerebral cortex.

15 I prefer Fuster's term 'active memory'[R2,3] on the basis that what is happening is that an engram in posterior cortex is being actively maintained. But working memory is what everyone currently calls it (including him).

16 For a detailed description and illustration of this processing, see section 11.5.2 in R13.

References

R1. Balsters, JH, Cussans, E, Diedrichsen, J, Phillips, KA, Preuss, TM, Rilling, JK, & Ramnani, N, Evolution of the cerebellar cortex: The selective expansion of prefrontal-projecting cerebellar lobules. *Neuroimage*, 2010. **49**: 2045–52.

R2. Fuster, JM, *Memory in the cerebral cortex*. 1995, Cambridge: MIT Press.

R3. Fuster, JM, Overview of prefrontal functions: E Pluribus Unum – Coordinating new sequences of purposeful action, in *The prefrontal cortex* (5th ed.), JM Fuster, Editor. 2015, San Diego: Academic Press. p. 375–425.

R4. Gaze, RM & Keating, MJ, The visual system and "neuronal specificity". *Nature*, 1972. **237**: 375–8.

R5. Goldman-Rakic, PS, Working memory and the mind. *Scientific American*, 1992. **267**: 73–9.

R6. Grove, EA & Fukuchi-Shimogori, T, Generating the cerebral cortical area map. *Annual Review of Neuroscience*, 2003. **26**: 355–80.

R7. LeDoux, JE, Emotion, memory and the brain. *Scientific American*, 1994. **270, June**: 50–9.

R8. LeDoux, JE, Rethinking the emotional brain. *Neuron*, 2012. **73**: 653–76.

R9. Loonen, AJM & Ivanova, SA, Evolution of circuits regulating pleasure and happiness with the habenula in control. *CNS Spectrums*, 2019. **24**: 233–8.

R10. MacLean, PD, Triune brain, in *Comparative neuroscience and neurobiology*, LN Irwin, Editor. 1988, Boston: Birkhäuser Boston. p. 126–8.

R11. McLaughlin, T, Hindges, R, & O'Leary, DD, Regulation of axial patterning of the retina and its topographic mapping in the brain. *Current Opinion in Neurobiology*, 2003. **13**: 57–69.

R12. McNaughton, N & Gray, JA, *The neuropsychology of anxiety: An enquiry into the functions of the septo-hippocampal system* (3rd ed.). 2024, Oxford: Oxford University Press. https://global.oup.com/academic/product/the-neuropsychology-of-anxiety-9780198843313?cc=es&lang=en&.

R13. Meyer, RL & Sperry, RW, Tests for neuroplasticity in the anuran retinotectal system. *Experimental Neurology*, 1973. **40**: 525–39.
R14. Moroz, LL & Romanova, DY, Alternative neural systems: What is a neuron? (Ctenophores, sponges and placozoans). *Frontiers in Cell and Developmental Biology*, 2022. **10**: 1071961.
R15. Murakami, Y, Uchida, K, Rijli, FM, & Kuratani, S, Evolution of the brain developmental plan: Insights from agnathans. *Developmental Biology*, 2005. **280**: 249–59.
R16. MacLean, PD, The brain in relation to empathy and medical education. *The Journal of Nervous and Mental Disease*, 1967. **144**: 374–82.
R17. Nauta, WJH & Feirtag, M, Fundamental neuroanatomy. 1986, New York: W.H. Freeman & Co.
R18. Nikitin, MA, Romanova, DY, Borman, SI, & Moroz, LL, Amino acids integrate behaviors in nerveless placozoans. *Frontiers in Neuroscience*, 2023. **17**: 1125624.
R19. Pennisi, E, The simplest of slumbers. *Science*, 2021. **374**: 526–9.
R20. Romanova, DY, Smirnov, IV, Nikitin, MA, Kohn, AB, Borman, AI, Malyshev, AY, Balaban, PM, & Moroz, LL, Sodium action potentials in placozoa: Insights into behavioral integration and evolution of nerveless animals. *Biochemical and Biophysical Research Communications*, 2020. **532**: 120–6.
R21. Sacks, O, *The man who mistook his wife for a hat, and other clinical tales.* 1985, London: Picador.
R22. Sansom, SN & Livesey, FJ, Gradients in the brain: The control of the development of form and function in the cerebral cortex. *Cold Spring Harbor Perspectives in Biology*, 2009. **1**: a002519
R23. Stephan, H & Andy, OJ, Quantitative comparative neuroanatomy of primates: An attempt at a phylogenetic interpretation. *Annals of the New York Academy of Sciences*, 1969. **167**: 370–87.
R24. Subramanian, L, Calcagnotto, ME, & Paredes, MF, Cortical malformations: Lessons in human brain development. *Frontiers in Cell and Developmental Biology*, 2019. **13**: 576.
R25. Swanson, LW, Brain maps 4.0—Structure of the rat brain: An open access atlas with global nervous system nomenclature ontology and flatmaps. *Journal of Comparative Neurology*, 2018. **526**: 935–43.
R26. Wicht, H & Lacalli, TC, The nervous system of amphioxus: Structure, development, and evolutionary significance. *Canadian Journal of Zoology*, 2005. **83**: 122–50.

5 What is an emotion?

'When I use a word,' Humpty Dumpty said in rather a scornful tone, 'It means just what I choose it to mean – neither more nor less.'
'The question is,' said Alice, 'whether you can make words mean so many different things.'
'The question is,' said Humpty Dumpty, 'which is to be Master - that's all.'
– Lewis Carroll: Through the Looking Glass

Emotion: the mind machine motor

Emotion is an enveloping word. It drags in perception, thought, feeling, motivation, and action. What exactly is emotion? Is it a single thing or are they a group of things? How does emotion relate to the rest of the mind? None of this is obvious. But the answers are important if, like me, you see emotion as the motor that drives the mind machine.

Subcortical survival circuits[1] operate somewhat like the knee-jerk or eyeblink reflexes. They make a simple, basic, contribution to explicit ongoing motivation or emotion. Neocortex can be viewed as dealing with colder, more cognitive calculations – resulting from and, in turn, changing motivation or emotion. It is the emotional limbic cortex that drives our ongoing actions and reactions and generates our mental disorders (Chapter 10).

But what *is* an emotion?

Why 'What is an emotion?' is a problem

You likely use the word 'emotion' (or emotion-related words) regularly; grumble when your emotions disturb you; or seek drug or psychological therapy when they bother you extremely. But it's often crucial in science to be very clear what a word means. Does what you want to study exist in the real world? You likely have a clear picture of a dragon (perhaps specifically Smaug or Drogon). But airline pilots do not need to look out for dragons.

'What is an emotion?' is a trap. It assumes that '*an* emotion' is a single real thing. I think the setting of this trap in 1884 by William James[R18]

DOI: 10.4324/9781003465027-5

was deliberate and a bit mischievous. Things have improved a bit in the last 140 years[R24,37] but more with the parts than with whole emotions; and there is still no consensus on the central issue of what *an* emotion *is*.

The main difficulties are semantic not scientific. We need to keep this clear, if science is to provide us with appropriate answers.[R28, pp. 2–3]

1 *Does the word only have one meaning in common usage?*
  e.g., a 'vest' (goes over the shirt in the United States; under the shirt in the UK)
2 *Does its referent exist?*
  e.g., a 'unicorn' (horn lying on a beach is a narwhal tooth)
3 *Does it have a distinct scientific meaning?*
  e.g., 'colours', the number of these is
  [physics] = infinite (wavelengths, some invisible)
  [TV/ monitor] = 3 (red, green, blue – the RGB[2] colour code)
  [folk usage] = 7 (in a rainbow) or more (e.g., turquoise)
4 *Does science have to make sense in English?*
  e.g., an 'electron' is a particle, or a wave, or both, or neither, all at once
5 *Do we agree what 'emotion' means?*
  The meaning of specific emotion words varies from person to person.[R11]
  The more general terms 'emotion' and 'affect' also mean different things to different people.[R28]

But if we have no agreed definition, how can we analyse 'it' scientifically and understand our own emotions in practice? Indeed, as we will see, 'an emotion' may refer to a cluster of things rather than a single process.

The answer is to use a biopsychological lens to look at, and integrate, the assortment of things that people have included in their concepts of emotion. Then see how they might fit together. I think you will find that you can *then* understand emotion, and *then* define it retrospectively. This is how 'electron' and 'atom' got defined (and will be updated when there is new physics). We can do the same for emotion labels about which there is disagreement, such as 'anxiety'.[R29]

Let's take 'emotion' apart

Is emotion a single thing or a pile of pieces? This is biopsychology, so we will get much of our answer from non-humans (particularly rats). Do they have emotions like ours, or minds at all? Are they just stimulus-response machines with no thinking in the middle? Surprisingly, experiments with rats are the easiest way to show that our emotions have parts, including cognitions and feelings,[3] and to determine their nature.

What things could be parts of 'emotion'? A quick Google of 'Emotion meaning in English' could equate it with: feeling/s: a conscious reaction to events/objects; unconscious reactions; and a cause of bodily changes and behavioural actions. Emotion is strongly linked, but not identical, to motivation – most definitions of emotion ignore this.

If emotion involves *reaction*, it must include an evaluation of objects or events – cognitions. These cognitions signal changed functional requirements. For example, if you detect threat, you need to temporarily increase your blood clotting capacity and your motivation to escape. For efficient escape, you need a range of bodily changes to, for example, maximize muscle performance. Suitable facial expressions and body gestures can also help with survival. So, all your effector systems can be involved: skeletal (action), skeletal (expression), autonomic (e.g., heart rate) and hormonal (e.g., adrenaline). We would normally label this complex package of reactions to an immediate threat as 'fear'.

Importantly, separate aspects of an object can be evaluated by separate sensory systems.[4] So, different aspects of the object can produce separate and independent reactions. We *see* these as parts of the same emotion because, under normal circumstances, they always occur together. But their separation is not surprising. Distinct changes (blood clotting/escape) can have distinct functions (responding to damage/preventing damage). So, evolution of one change (e.g., blood clotting) does not change the adaptive value for evolution of the other (e.g., escape). Multiple positive feedback loops can also link the different components. Thinking fearful thoughts will change your facial expression and increase adrenaline levels; and facial expression (see Experiment 5.1) and adrenaline in your bloodstream can, in turn, increase your thoughts of fear.

So, what binds a set of thoughts, motives, feelings, and actions together into an emotion is that they result from *a specific functional class* of recurring situations. They, and their interactions, share a common evolutionary adaptive history. They all cope with the same type of recurring situation – in that sense, they have a common function. Even unpleasant emotions (anxiety, depression, pain) are normally adaptive. Pain warns you where you just put your hand when cooking. It is only when they occur inappropriately or excessively that they represent disorder.

Emotion and cognition

Simple stimuli – events – can seem to trigger emotion. But even rats must interpret a stimulus (e.g., a cat) not just detect it (similarly, when *you* meet a snake). Often in biopsychology, we must compare several conditions to draw a conclusion. This is one of those times. It's complicated. Feel free to skip to the next section if you want to just trust me. But read on if you want evidence that rats draw conclusions.

First, present a rat with a moderate intensity tone and measure its CORT[5] hormone levels and behaviour. You will see no increase in CORT nor an increase or decrease in behaviour. Second, present a shock. There will be no increase in CORT[R4] but there will be an increase in active behaviour (escape, vocalization).

Finally, pair the tone with an unavoidable shock several times and then present the tone by itself. There will be an increase in CORT[R3] and a *decrease* in active behaviour (via freezing or conditioned suppression). Neither the tone nor the shock previously produced these effects. Clearly, CORT and motor inhibition are the result of the rat now anticipating a shock (and being anxious about the threat). It cannot be simple stimulus-response linking. An internal system is linking particular cognitions with appropriate evolved reactions.

But what if the shock is avoidable? Initially, we pair tone and shock as before. *While* the rat is learning to make an avoidance response, CORT levels are high (consistent with the unavoidable case). But then, as avoidance becomes more successful, CORT drops down and, when avoidance is completely successful, returns to pre-shock levels.[R9] Interestingly, if we then block the avoidance response, but with no shock delivery, CORT returns. Again, clearly, CORT release depends on whether the rat is COGNITIVELY anticipating shock – not on the precise arrangement of stimuli. Note that there is no change in CORT during equivalent learning with food rather than shock.[R8]

So, even in supposedly simpler rats, emotions result from *assessments* of situations and resultant goal-directed actions.[6] They are not simple reactions to simple stimuli. Of course, when dealing with humans and their expanded neocortex (Chapter 4), the cognitions can be more complicated and look further forward in time. But the link between cognition and emotion is not different.

Emotion and external action

In its original Latin form (emovere), 'emotion' includes the ideas of moving oneself away, or removing or dislodging something. That is, they are active external responses to what you are perceiving. An emotion involves actions (or action tendencies) with functional results[7]: escape from threat; consumption of food when hungry; courting of a potential mate.

Surprisingly, emotional 'action' can involve stillness. A motionless rat can be 'freezing' to an immediate threat, showing 'defensive quiescence' to a possible threat, and, in females, showing 'lordosis' in response to the presence of an attractive male. You can tell these apart immediately from a still photo. Freezing involves a tense, bunched-up posture – ready for explosive escape or jumping attack. Defensive quiescence involves a low, stretched-out posture, ready for fast backward movement. (Imagine a rat easing out of its burrow while checking for the cat.) Lordosis involves the opposite back curvature to freezing. The female lifts her rear end to facilitate mating. As you can see, these 'stationary actions' are all functional in their specific context – increasing the chances of survival or reproduction.

Emotion and external expression

Expressions, particularly facial expressions, are emotion-related actions. But it's harder to see why they arose compared to bodily expressions. You could wave a wand and say "oh, they are just there for communication". But Darwin would not be happy with this idea:

> The view that [people] had been created with certain muscles specially adapted for the expression of [their] feelings struck me as unsatisfactory. It seemed probably that the habit of expressing our feelings by certain movements, though now rendered innate, had in some way been gradually acquired.
>
> (Darwin[R10, p. 19])

Emotional actions (stationary or otherwise) provide information about intentions. – even before they are expressions in the conventional sense. They are signals. You see them and you may change your action. Signalling value can then evolve – beyond the original function.

If you are planning on biting someone, and don't want to bite your own lips, it's best to bare your teeth. This is best automatic and, so, quick. As you can see from Figure 5.1, the human version of the grimace is more like ∞ than ⊂⊃. In the mandrill, ∞ displays large canine teeth (that we no longer have). The ∞ shape was retained by us for its signalling value. The resultant evolved 'grimace'[8] has a specific form, using specific muscles. These muscles differ from the chimpanzee equivalent of the smile (⊂⊃ the 'play face'). For the human play face, Google 'baby smile'. Note the relaxed oval open mouth. Our smile uses the play face muscles.

So, we have an evolutionary sequence each step creating pressure for the next: reproduce 🡐 survive 🡐 win 🡐 attack 🡐 prepare to attack 🡐 signal intention (maybe avoiding need to attack) 🡐 bluff. The last three of these can

Figure 5.1 The 'grimace' – getting ready to bite? (A) Small girl in a rage. (B) Kabuki actor portraying rage. (C) Mandrill showing threat display. The mandrill (and anthropoid apes) draws down the corners of its mouth, so that the full extent of the upper canine tooth is visible. Human beings do the same, although our upper canines are now nothing like as long. Figure and legend modified from Eibl-Eibesfeldt.[R13, p. 18] Used with permission of Taylor & Francis Group LLC – Books; permission conveyed through Copyright Clearance Center, Inc.

be shaped by a need for a specific communicative value (in the case of bluff, a false value from the receiver's point of view).

That said, some facial expressions appear to retain a basic function without adding value in terms of communication (Figure 5.2). For example, *expressing* fear in your face (eyes and nose open) enhances the acquisition of sensory information. Why would you want to signal this to a predator? In contrast, *expressing* disgust in your face (eyes and nose closed) suppresses acquisition. Usual causes of disgust would not be noticed or cared. The details of these are particularly easy to explain in evolutionary terms – unlike sexual displays which may have completely arbitrary origins. They appear to prepare the person for appropriate action; and, in this case, fit Darwin's suggestion that opposite functions will produce opposite forms of expression[9]:

> Some theorists consider emotional expressions as direct expressions of internal states, implying that they are involuntary, cannot be controlled and are inherently honest. Others see them as more intentional and/or as indicators of the actor's future behavior. ... these different viewpoints are not mutually exclusive. Recent insights indicate that, in both apes and humans, some emotional expressions can be controlled or regulated voluntarily, including in the presence of audiences, suggesting modulation by cognitive processes. However, even non-intentional expressions such as pupil dilation can nevertheless inform others and influence future behavior. In sum, while showing deep evolutionary homologies across closely related species, emotional expressions show relevant species variation.
>
> (Kret et al.[R22, p. 378])

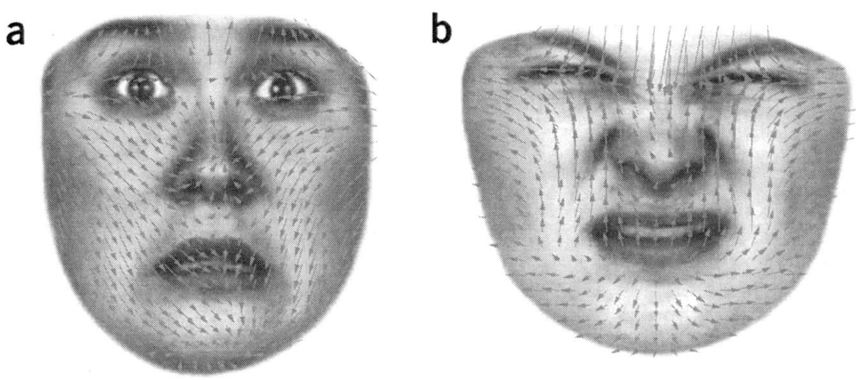

Figure 5.2 Opposite functional expressions. (A) FEAR: eyes and nose open – you see more; move your eyes faster; take in more air without extra effort; maybe smell more. (B) DISGUST: eyes and nose closed – you see less; move your eyes slower; smell less. Figure from Susskind,[R42] used with permission of Springer Nature BV; permission conveyed through Copyright Clearance Center, Inc.

Emotion and internal action

A more superficially mysterious aspect of emotions is their noticeable internal bodily changes. We often talk about 'feeling' an emotion. But the internal changes are driven by the physiological needs of our external emotional actions. Indeed, many bodily changes during emotion are not felt (e.g., changes in blood clotting). The ones that are felt occur for purposes unconnected with the associated feelings (e.g., adrenaline to reverse the effects of fatigue). That is, it's best to think of bodily changes as under the control of an internal 'motor' system. Different patterns of internal response evolved to fulfil different adaptive bodily functions – our feelings are secondary.

This is a very old idea. Nearly 90 years ago, Walter Cannon pointed out that "bodily changes in pain, hunger, fear and rage" shift organs into overdrive, ready for action, and then shut them down to conserve resources when action is not required. So, internal changes can be understood in terms of physiological rather than psychological function (albeit often having the psychological consequences we call feelings). In functional terms, Cannon itemized:

> increased blood sugar as a source of muscular energy… increased (adrenaline) in the blood as an antidote to the effects of fatigue… the vascular changes… favourable to supreme muscular exertion… the value of increased red blood corpuscles… the changes in respiratory function (as) favourable to great effort… the utility of rapid coagulation in preventing the loss of blood.
>
> (Cannon[R6, p. xiv])

You are likely to notice increased breathing and heart rate (and to have read sentences like "His pulse quickened at the thought of seeing her again", https://www.merriam-webster.com/sentences/quicken). But what about changes in your red blood corpuscles or blood coagulation? The 'why' of the changes that you do notice is their adaptive function. Feelings are only mysterious if you seek a psychological explanation for them.

Bodily changes are controlled by two distinct nervous systems (Figure 5.3). The *sympathetic* does 'go'. It produces changes as needed. The *parasympathetic* returns things to rest. You need both. Change is only briefly adaptive. Blood coagulation that is always high will generate clots. You will die from a stroke or heart attack. Conversely, with haemophilia or anticoagulant medications, you can bleed to death. Switching up when you expect to get bitten and then down when you are safe gets you the best of both worlds.[10] You will live longer, reproduce more.

Emotion and feelings

Try Googling 'Emotion meaning in English'. All the major dictionary definitions link emotions with strong feelings. The feelings, in turn, are often seen

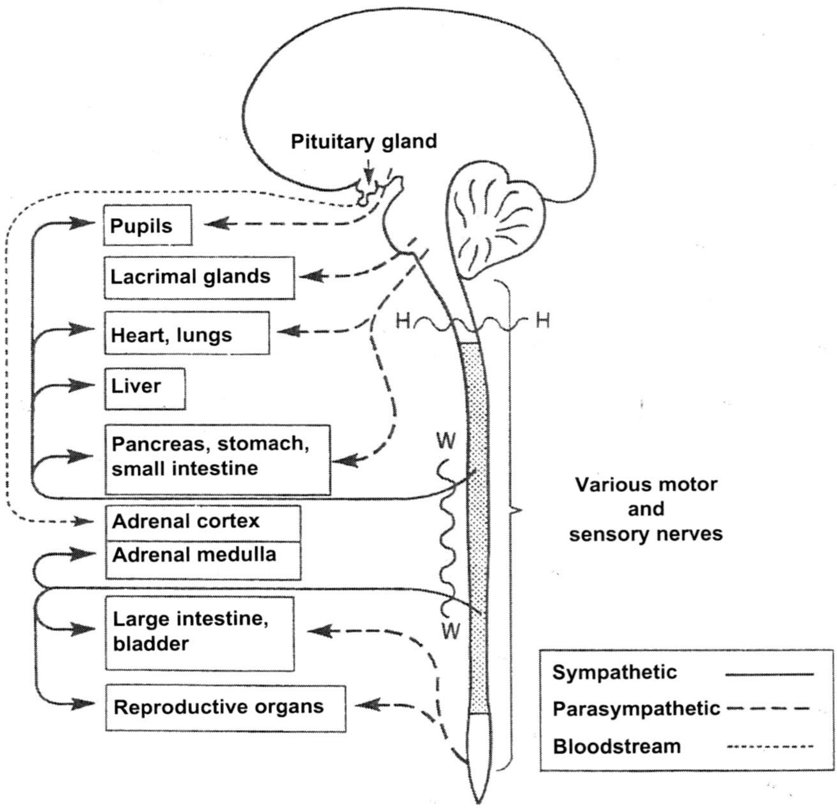

Figure 5.3 Cross-sectional view of the brain and spinal cord showing the outflow of the autonomic nervous system. This is divided into two parts: the sympathetic and the parasympathetic. They target most of the organ systems of the body with some variation for each division. The sympathetic system can be thought of as a 'go' system; the parasympathetic as 'return to rest'. The sympathetic division derives mostly from the middle portion of the spinal cord and so can be selectively cut (W-W) or variably damaged in addition to sensory and motor systems in paraplegics (H-H) with results described in the text. Figure and adapted legend from McNaughto. [R28] Reproduced with permission of Cambridge University Press through PLSclear.

as the perception of bodily reactions. This raises a chicken-and-egg question: Does the emotion produce the bodily changes that produce the feelings; or, do the bodily changes produce the emotion and its feelings? This is an old debate in its either/or form that is best resolved with a picture of parallel, but interacting, processes.

The debate is complicated by some using 'feeling afraid' to mean 'perception of threat'; while, for others, 'feelings of fear' are 'perception of the bodily reaction to threat'. As with 'emotion' in general, I will use biopsychology for clarity.

A mutation occurred in your distant ancestors allowing detection of a predator (e.g., of the shadow as it swam over). If your ancestor escaped (and so *could be* an ancestor), they could then learn to avoid. An additional mutation (in the circuit that did the learning) could then generate an innate avoidance reaction. Similarly, further mutations (in autonomic circuits) could increase the speed of escape and avoidance. Now, if you met a bear (and had been taught that it is dangerous) you could be afraid *and* feel fearful.

Your primate ancestors took a next step with a particularly troublesome predator. Snakes could easily attack them even in trees. As a result, they, and you, have an innate snake detector (in the superior colliculus and then thalamus) that means you don't have to *learn* a snake is dangerous. This detector triggers arousal and orientation (in the amygdala) that then generate autonomic and skeletal responses (via the medial hypothalamus and periaqueductal grey).[R19]

At least some of these autonomic responses will be perceived as sensations. It could be that that is all they are there for – with skeletal behaviour being the important element of emotion. But, even for Darwin cited in Mandler[R25, p.47], without adaptive autonomic changes you would not have true emotion in the everyday sense. So, a common view of emotion is that you see the bear, immediately become afraid and then, with some delay in release of the relevant hormones, feel the emotion.

William James and, independently, Carl Lange suggested the opposite sequence: that bodily sensations necessarily occur before emotion proper. "Without the bodily states following on the perception …we might see the bear and judge it best to run, … but we could not actually *feel* afraid".[R18, p. 19 of reprint] In one sense, this is a tautology – if feelings *are* perception of bodily state, it follows that, without the bodily change, you cannot have the bodily sensations. But James meant it more strongly – that the mental aspects of fear are generated by the bodily changes. In this sense, it could be wrong. If feelings result from autonomic changes (each with its own adaptive history) then they are likely to be in parallel with adaptive external actions and expressions – not the cause of them.

Cannon emphasized the independent adaptive values of autonomic changes. Likely because of this, he disagreed strongly with James's view. His main objections were that

> 1. total separation of the viscera from the central nervous system does not alter emotional behaviour. [For example] all superficial signs of rage were manifested … hissing, growling, retraction of the ears, showing of the teeth…. 2. The same visceral changes occur in very different emotional states and in non-emotional states. … 4. Visceral changes are too slow … the organic changes could not occur soon enough to

> be the occasion for the appearance of affective states. ... 5. Artificial induction of the visceral changes typical of strong emotions does not produce them ... [although some cases] remarked 'I feel as if afraid', 'as if awaiting a great joy' [etc.].
>
> (Cannon[R5])

With his student Philip Bard,[11] Cannon developed the theory that emotional reactions

> reside in subcortical centers ... ready for instant and vigorous discharge when they are released from cortical restraint and are properly stimulated. ... *the peculiar quality of the emotion is added to simple sensation when ... thalamic processes are aroused.*
>
> (Cannon[R5, pp. 19–20])

Bodily changes and skeletal outputs occur almost simultaneously with the thalamic changes leading to James's mistaken picture of their role. Cannon and Bard's key data came from non-human experiments. However, Cannon also noted that human paraplegics reported no changes in their emotions despite loss of autonomic changes – and as we shall see this may not strictly be the case.

Effects of feelings on emotions

George Hohmann was a paraplegic, serving as a Staff Psychologist in the Spinal Cord Injury Service, and had extensive experience working with paraplegic and quadriplegic patients. From his personal experience and his patients' clinical reports, he and the Chief of the unit thought that reports of no change in emotions in paraplegics were suspect. He undertook structured interviews of five groups with different levels of spinal injury. He thought paraplegics would under-report subjective changes to intact persons so as not to be judged 'abnormal'. He found, indeed, that some subjects previously tested by intact psychologists admitted giving different replies to him on the grounds that he would understand their position better.

From the interviews, he concluded that

1 Individuals with spinal cord lesions report significant decreases in experienced emotional feelings associated with sexual excitement, anger, fear, and an overall estimate of emotional feeling, as compared to those experienced before injury.
2 Such subjects report a significant increase in emotional feelings related to sentimentality.
3 The data further indicate that, in spite of decreases in some emotional feelings, overt emotional behaviour may continue to be displayed".[R17, p. 155]

These are intermediate results with 'decreases' and 'increases' rather than presence or absence. (In explaining increased sentimentality, note that crying is likely to be a common feature of sentimentality and is controlled by sympathetic output that would not have been damaged, see Figure 5.3.) Hohmann's results appear to both partially support and partially contradict both James and Cannon. The results could also have been biased by the situation and so should be treated with caution – as should some patients' comments, which are nonetheless interesting in departing from the way we normally talk about emotions and feelings:

> Sometimes I *act* angry when I see some injustice, I yell and cuss and raise hell, because if you don't do it sometimes, I've *learned* people will take advantage of you, but it just doesn't have the heat it used to. It's a mental kind of anger.
>
> (Hohmann[R17, p.154])

> I used to have a hot temper but now it's nothing compared to what it was. Now I get kinda mad one minute, and the next minute it'll be like nothing happened. Seems like I get thinking mad, not shaking mad, and that's a lot different.
>
> (Hohmann[R17, p. 151])

We can account for this, in part, if Cannon-Bard is right. As in the snake example, above, where we have a good idea of the circuitry, a specific object is detected and activates the thalamus (and potentially neocortex), which then generates appropriate skeletal and autonomic responses. Lacking an autonomic output, one would still have – on a millisecond time scale – appropriate emotional behaviour (e.g., escape) and, according to Cannon, emotional 'feelings' resulting from the thalamic input to the cortex. But this would lack the sensations of sympathetic activation and be 'a mental kind of anger'. Adding the sympathetic reaction back in would add in bodily sensations but on the slow time-scale noted by Cannon. This should, at least, add 'heat' allowing one to become 'shaking mad'. However, it also appears from the reports to increase the duration and possibly intensity of the emotion being experienced.

This idea fits with the typical report 'I feel as if I am afraid' from people given adrenaline injections; but it also fits with the fact (as Cannon notes) that, in those already predisposed to an emotional reaction, injection of adrenaline can elicit the relevant emotion, including panic attacks.[R23] Similarly, high doses of caffeine can result in anxiety and panic (especially in those who are already panic-prone).

I won't recommend a self-experiment on this for safety reasons – but you can likely find observational examples of this in your daily life. The important points are that artificial 'autonomic' changes can, fairly non-specifically, support emotion and that this is "only when an emotional mood already exists".[R5, p. 114] So, when a normal emotion elicits its normal autonomic changes, this should amplify both intensity and duration of the emotional reactions.

Effects of skeletal change on emotion

If internal action is a result of an activated thalamus, with the activation then feeding back to amplify emotion, what of external action and facial expression? Cannon agreed with James "that the assumption of an attitude does in fact help to establish the emotional state which the attitude expresses".[R5, p. 121] That is, sit in a moping posture and you will tend to be sad; sit upright and you will tend to feel happy. But Cannon argued that this result was essentially an 'I feel as if' change not true emotion itself. If action is like feeling, then, it is a product of emotion that can then feedback to change emotion – with opposite postures producing opposite effects (which is why soldiers are taught to stand tall). The effect of posture is something you can try and test for yourself at home but the effects will be small – and it's impossible to know if any emotional change is due to the posture or your expectations of it.

Related feedback from facial expressions has also been suggested to enhance or even produce emotion (the 'Facial Feedback Hypothesis'). Facial expressions of emotion are consistently linked to specific patterns of muscular contraction and at least seven are recognized across cultures (anger, fear, surprise, sadness, disgust, happiness, and contempt).[R14]

This allows you to 'talk' people's faces muscle-by-muscle into basic emotional expressions without naming the emotion. You can then ask them if they have any emotional feelings and, if so, what they are like. As an example of how you can go about this, try Experiment 5.1 as a party trick (but don't take the design or results too seriously). I originally tried this in a class just as a demonstration of the procedures and was surprised to get the predicted results, so I repeated it over several years varying the emotional expression (see the experiment box for my results). Note that this experiment could be producing 'I feel as if' reports not real emotion.

More convincing results are obtained when facial expression is combined with emotion-inducing slides. Importantly, this allows for match or mismatch between the facial emotion and the slide-induced emotion. Rutledge and Hupka[R38] took considerable pains to hide from their participants the fact that they were studying emotional expressions and used a questionnaire covering ten emotions but only attempted to manipulate two (joy and anger). They combined the two emotional expressions with five types of slides: high joy, low joy, neutral (a blue screen), low anger, and high anger. The selection of slides was based on the relevant emotional scores in a pilot experiment. This produced an essentially linear change in relevant emotional intensity scores, with low anger and high joy to high joy slides and high anger and low joy to high anger slides. Critically, they found (Figure 5.4) that "subjects reported feeling more angry and less joyous when in an anger pose and, similarly, more joyous and less angry when in a joy pose". This effect of facial expression was consistent across the emotion-inducing slide types.

Experiment 5.1 Falsifying the Cannon-Bard theory – facial feedback

This experiment is best carried out as a party trick as you need ~20 volunteers. Give each a pen and a piece of paper that says the following:

> Look straight ahead and, especially, don't look at your neighbour while you are following the instructions that I will read out. Then HOLD THE EXPRESSION UNTIL I TELL YOU TO STOP

Then, write down your feelings while you were holding the expression

----

Circle the emotion that the feelings most remind you of or felt like
1. Anger 2. Disgust 3. Fear 4. Happiness 5. Sadness 6. Surprise

Once you have handed out the pieces of paper, explain that you will talk their faces into a particular expression, ask them to hold it for 30 seconds, and then ask them to write down any feelings they had. Tell them it's important for them not to look at each other so they can't see their expressions. Then read out one of the instruction sets provided below. Collect the responses, making sure they have circled an emotion, and tally the results. The percentages I got in a lecture class (intended emotion highlighted) are:

	Happiness	*Disgust*	*Anger*	*Surprise*	*Fear*	*Sadness*
1998–2001	6.8	13.6	72.1	1.4	2.0	4.1
2002–2005	42.3	17.2	16.0	14.7	2.5	7.4
2007–2010	7.0	3.2	1.3	70.9	5.7	12.0
2011–2014	2.0	17.6	56.3	4.5	8.5	11.1

Instruction set 1: 'ANGER' adapted from Pinel[R32, p. 437]:

- Pull your eyebrows down and together;
- Raise your upper eyelids;

- Tighten your lower eyelids;
- Narrow your lips;
- Press them together.

HOLD THAT UNTIL I TELL YOU TO STOP (now time 30 s)
NOW PLEASE WRITE DOWN ANY FEELINGS YOU HAD

Instruction set 2: 'JOY/HAPPINESS' adapted from Rutledge and Hupka[R38, p. 228]**:**

- Open your mouth slightly;
- Tighten your cheek muscles (they are a bit below your eyes);
- Keep your mouth slightly open;
- Move the corners of your mouth up and back;
- Hold all of these and raise your lower eyelids slightly.

HOLD THAT UNTIL I TELL YOU TO STOP (now time 30 s)
NOW PLEASE WRITE DOWN ANY FEELINGS YOU HAD

Instruction set 3: 'SURPRISE' adapted from Pinel[R32, p. 436]**:**

- Pull your eyebrows up so as to wrinkle your forehead;
- Open your eyes wide (white should show above the iris);
- Slacken the muscles around your mouth;
- Drop your jaw.

HOLD THAT UNTIL I TELL YOU TO STOP (now time 30 s)
NOW PLEASE WRITE DOWN ANY FEELINGS YOU HAD

Note that the intensity of emotion elicited by the slides themselves is not extreme (30% of maximum) and so a larger increase could have been measured. As it is, the effect of the two expressions is similar in size and produces an increase in intensity of only about 4%. This weak (potentially 'I feel as if') supplementary effect is consistent with supplementary effects seen with autonomic changes.

Separating the parts – Separation Anxiety

As you have seen emotion has a lot of parts, which show some interactions. We now need to look at how those parts combine to make 'an emotion' and how far they may be separate, operating in parallel, rather than all the result of the same cause. I'll use, as a paradigm example, Separation Anxiety.

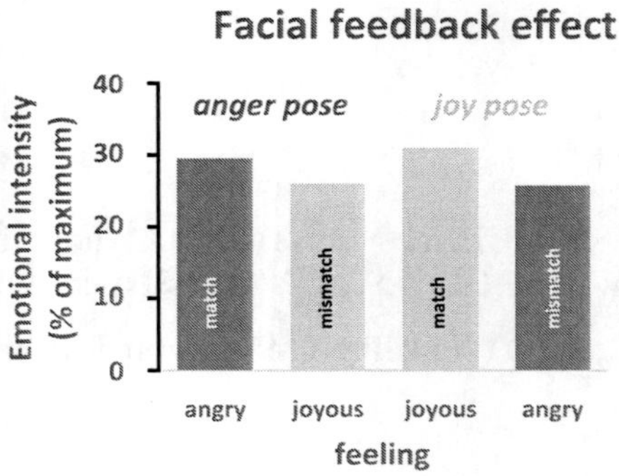

Figure 5.4 Effects of posed facial expression on slide-induced emotion. The values plotted are emotional intensity as a percentage of the maximum scale score. A match between slide emotion and expression emotion produces a higher intensity than a mismatch. Data taken from Rutledge and Hupka[R38] and converted to percentage of scale maximum.

Separation, or the possibility of separation, from home or from people to whom they are strongly attached, can make people, particularly children, anxious. This Separation Anxiety is often adaptive, reducing the risk of harm. As with many other adaptive reactions, it can be classified as a maladaptive disorder if it occurs at an inappropriate stage of development or is otherwise excessive.[12] It occurs, and in much the same way, in other primates; and in rats – who we can more easily use in controlled experiments.

Myron Hofer[R16] removed rat pups from their mother when they were two weeks old and observed Separation Anxiety: decreased heart rate and increased locomotion, grooming, defecation, and urination. Removal of the mother necessarily removed their source of food, warmth, and behavioural support. Which of these caused the observed emotional changes?

Hofer found that the heart rate decrease was not reversed by providing warmth, four-hourly feeds with milk, or frequent feeding with a non-nutritive solution. It was reversed (via a neural rather than metabolic route) by frequent feeding with milk. Conversely, the increased locomotion, grooming, defecation, and urination were not reversed by frequent feeding with milk. But these behavioural changes were reversed by the presence of a non-lactating foster mother rat. So, the heart rate and behavioural changes are independent of each other.

Putting the parts back together – 'What is an emotion' revisited

External action, external expression, and internal action are elicited by the same situation (such as a predator). But, as we saw earlier, each has its own adaptive function. The Separation Anxiety results show that these different adaptive functions can result in independent neural control.

Importantly, the regular co-occurrence of the different components of Separation Anxiety is not caused by some single common neural trigger. It is caused by the necessary coincidence of the different eliciting stimuli. Separation from the mother (an evolutionarily common situation) cannot be done in parts. Separation removes all the different stimuli (milk and cuddles), generating the parallel autonomic and behavioural reactions. So, the consistent pattern of changes, which we recognize and label as the emotion, results from a specific class of situation that has recurred in evolution not from a specific central control mechanism.

It follows that we can define emotions via the adaptive value that drove the evolution of their parts (Table 5.1). It's simplest to think of this as their 'purpose' but it's important, as emphasized by Richard Dawkins,[R12] to remember that this is a historical (technically 'teleonomic'[R28]) not future-oriented (teleological) concept. The explanation of what we see currently (e.g., the human grimace) may involve a history of 'purposes', each following on from the previous one. At no time during this history was the current version of the emotion 'intended' as the end point.

If each part has its own adaptive equation, there are two ways each part could be controlled. In the case of snakes discussed above, a single innate mechanism detects the snake and activates threat detection, which is likely shared by other dangers such as spiders. This threat detection can then trigger behaviour and autonomic reactions. In the case of separation, separate mechanisms triggered behaviour and autonomic reactions. But with some cases of separate control there could have been times when only one of the two reactions was triggered. This would create adaptive value in linking it with the other so as to generate or enhance its response. Separation of processes would explain 'I feel as if' responses. Later linking would explain the weak component interactions in some emotions. These can operate in both directions.

Table 5.1 Emotions defined via teleonomic purpose and 'adaptive value'

Emotion	*Teleonomic purpose/goal/ function*	*'Adaptive value'*	
Fear	Repulsion from danger	Survive threat	[➔ offspring]
Anxiety	Cautious approach to danger	Get food despite threat	[➔ offspring]
Lust	Copulation	Offspring	
Love	Pair bonding	Offspring survive through care	
etc.			

Note that, for example, we would expect love to be strong in albatrosses (see Chapter 2) as a chick needs food supplied by both parents if it is to survive.

Table and legend adapted from McNaughton.[R30]

'An emotion', then, is a *set* of evolved reactions. They are bound together by the nature of the recurring type of situation (e.g., threat) to which they adapted. But they can be more or less independent of each other in control terms. Where members of the set interact with positive feedback, each amplifying the other, there is the potential for not only maintenance of the emotion (as with Hohman's patients prior to their injury) but also explosive reactions like panic (as with injections of adrenaline in a threatening situation).

So, an emotion is a kind of mental stew of mixed ingredients (Figure 5.5). Like any specific stew, the recipe can vary from time to time. Also, like different types of stew (Irish, goulash, etc.), we can recognize and name many distinct types of emotion. Nonetheless, taken one at a time, we can

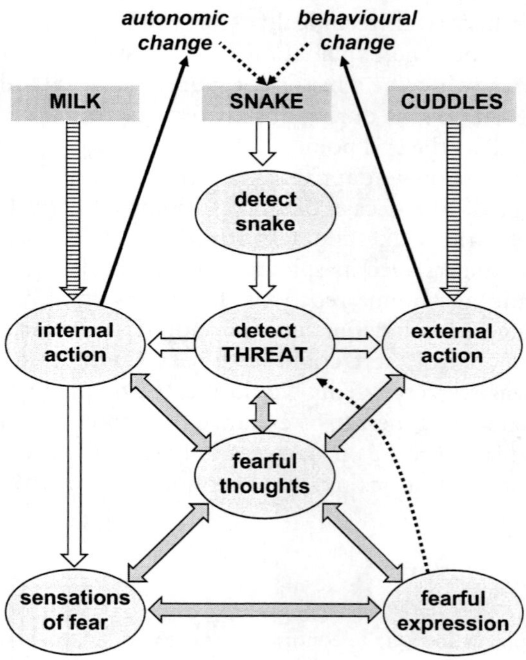

Figure 5.5 **What is an emotion?** – the mental mechanics. An emotion involves to a greater or lesser extent all the elements connected by the grey arrows. The **innate** triggers (a snake; the milk+cuddles elements of a mother) are shown activating (white arrows) and suppressing (cross-hatched arrows) the emotion, respectively. Triggers can also be learned. The emotion progresses through the combination and interaction of the elements – with internal and external action generating outputs that feed back to the original causes of the emotion by changing the internal or external environment (dashed arrows). Some expressions (e.g., fear, disgust) can change the reaction to the eliciting stimuli by altering the level of sensory input (up and down, respectively). The changes in a specific emotion are linked by their parallel evolution in response to a common recurring functional requirement (Table 5.1).

understand the ingredients individually. Importantly, if we hold in mind the functional nature of their common eliciting situation, we can understand why the emotion has the specific combination of ingredients that it does. While we have focused here on innate stimuli that can activate emotions, as we will see in the next chapter, learning can replace these by previously neutral stimuli. So, while much of the content of an emotion is innate, its triggers need not be.

Emotion blends

An extra complication for our scientific understanding of emotions is that the basic ones can be elicited in parallel, producing blends. Our perception and naming of emotions can then be much more complex than the limited number of basic emotions. This is like our perception and naming of colours (such as turquoise) is more complex than the three fundamental colours detected by our light receptors and projected from TV screens.

Figure 5.6 shows the blending of an increasing tendency to fight with an increasing tendency to flee. The different expressions are given different names by ethologists; and the combined expression (i) may reflect an internal state that is given a different name than the separate expressions (c, g). But, at least in this particular case, the expressions can all be seen as occupying a two-dimensional (fight/flight) emotional space.

Emotion and motivation

Emotions produce sensations and urges. But they also motivate specific actions, often with major impact on others. Figure 5.5 glosses over the 'why' of actions. Fight and flight are not necessary results of 'detect threat' and 'fearful thoughts'. Conversely, you can surely choose to leave a potentially dangerous situation (or look right and left before crossing a road) without *feeling* fear?

So, do you have to feel an emotion to be motivated to produce an action? Or, indeed, are emotion and motivation completely separate? This question is particularly important when we look at motivated eating.

Suppose you 'feel hungry'. Does this involve emotion or just motivation? Obviously, one must eat to live. For this you just need to seek and then eat food. But a brief look at my waistline will tell you that I, at least, live to eat (with cooking my favourite hobby and my personal-tested-recipe book currently close to 400 recipes). To understand my waistline, and the current obesity epidemic, we need to understand emotional factors. For this, we need to understand the biology and psychology of hunger.

Surely, we eat to replace an internal deficiency – to make up the energy lost since our last meal? Unlike fear generated by the sudden appearance of a predator, no stimulus triggers hunger and eating. So, the simplest theory of both hunger and stable weight is the 'set-point' theory. Food control is

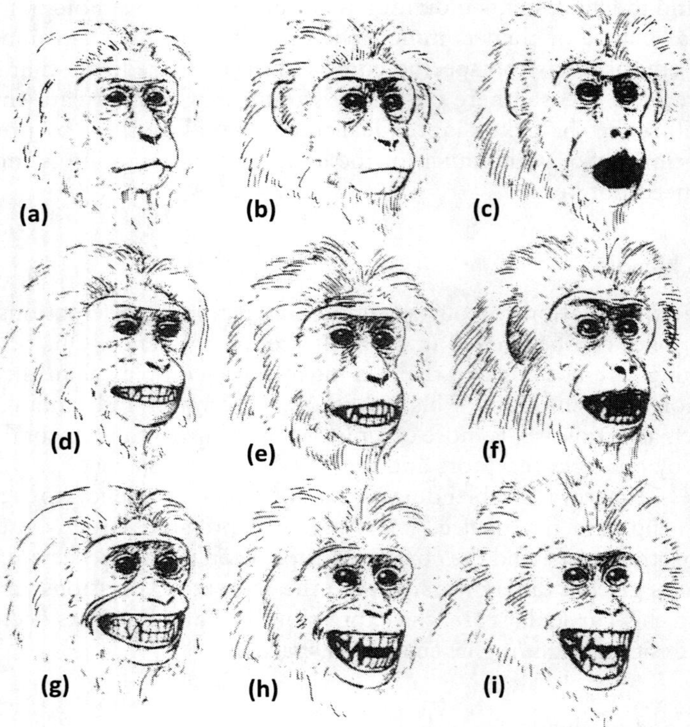

Figure 5.6 Blending of emotions. This is most easily seen with facial expressions. (a, b, c) Reflect an increasing tendency to fight. (a, d, g) Reflect and increasing tendency to flee. (a, e, i) Shows their combination. Figure and adapted legend from Chevalier-Skolnikoff.[R7] Copyright holder's address not available. Original artwork by Eric Stoelting.

like temperature control. You have a foodstat like a thermostat on a heater. For set-point theory, energy (e.g., glucose in the bloodstream) is measured (e.g., by a 'gluco-stat'). When energy levels drop below the set-point of the gluco-stat, you become hungry. The 'heater' is switched on – you search for food and eating. This will correct the energy level – eventually.

While simple and intuitive, the set-point theory does not fit the facts. Neither glucose levels nor hunger changes steadily between meals – glucose levels are held constant. Instead, insulin (which controls glucose levels[13]) is released a few minutes *before an expected* meal.[R40] This decreases glucose levels in anticipation of the metabolic effects of eating. Only then do you become hungry. This is why diabetics inject insulin just before a meal. Likewise, stable weight in the long term might be thought to involve a set-point for fat levels (a 'fat-stat'). But a range of issues[R41] – including weekend weight gain – do not fit with set-point theories.

In particular neither a gluco-stat nor fat-stat would explain why:

- eating stops well before any metabolism of the food can change fat or even glucose level
- eating is not inhibited by body fat
- calorie pre-loading does not reduce meal size

There are several current theories[14] of body weight and fatness regulation[R41] that can be lumped together as 'positive incentive' theories of hunger. Put simply, you often eat because of your current wants not because of your current needs. We need to look to evolution to understand why this is.

A key point is that famine occurs frequently (as I write, it's in Gaza and human-induced). With enough fat, you will survive. You will also do better during a hospital stay.[R39] So, your ancestors evolved to 'over-eat' now and survive the future; with annual weight gain in hibernating animals. But we need to explain how our weight is so stable (and hard to change when we want). Evolution can account for this, too.

You should keep weight down if you want to run away from predators[R41]; or want to limit your chances of heart or knee problems. Why add weight during the summer if you don't hibernate? (*We* tend to put on weight during the winter with increased eating.[R2]) The good/bad balance (too much fat/too little fat) matches Cannon's picture of internal emotional changes. Evolution has a lot of these balances.

So, we need to explain short-term starting and stopping of eating; *and* long-term stable weight control; and re-starting of eating using the 'pudding stomach' that most of us have available for extremely high-calory sweet things at the end of a meal despite being 'full'. Each of these control issues has its own explanation.

The most basic trigger for eating is meal time. In my own case, I don't feel hungry in the morning and breakfast is only a large cup of strong coffee. Despite my last food being the previous evening, I don't feel hungry until about 11:55 but I get grumpy if I cannot have lunch at 12 noon. (My wife has a more normal habit pattern and will have had breakfast and morning tea by then.) I then do not feel hungry again until just before my evening meal at 6:30 (when my wife will also have had afternoon tea). Linked to this habitual time trigger is release of a hormone, ghrelin, from the stomach about half an hour before a meal. This is thought to confirm that the stomach is ready to receive food.

But every now and then, our Psychology Department has a 10:00 'morning tea' as a celebration. (Set-point and metabolic theories don't account for the worldwide use of feasts as a means of celebrating.) I feel (1) a bit hungry in anticipation (maybe due to ghrelin); and (2) very hungry when I *see and smell* the small savoury pastries, fish bites, and small kebabs. I never (3) want a smaller lunch at 12:00 as a result. These effects have been studied as (1) pre-meal hunger (whenever the meal); (2) incentive motivation (the presence of tasty food makes you feel hungry); and (3) calorie pre-loading. I don't think I need to suggest experiments to you – I am sure that your memory will be enough.

An interesting aspect of incentive motivation is that it is taste-specific. As an illustrative experiment, eat just one specific food you like until you feel full. Keep all other food out of sight. Now bring out another food you like that tastes different, you will feel hungry again even if both are savoury or both are sweet. The classic 'pudding stomach' is similar but operating across classes of tastes (savoury/sweet) rather than single ones. One explanation for this taste-specificity is that it promotes a balanced diet in terms of nutrients by eating a range of food types for any fixed calory intake.

Taste is not the only stimulus that can initiate eating when sated. Rats can learn that a light + buzzer means meal delivery. Once trained, they will eat this meal, in response to the stimulus, even if they are sated. If you follow the meal with lithium chloride injections that induce sickness, this produces a learned taste aversion. Surprisingly, the light + buzzer still elicits food approach; and the rats start eating as soon as before. But then they eat much less (30%) than control rats who did not have taste aversion training. This shows that starting and stopping eating are controlled by quite separate mechanisms.[R44] Taste aversion is a great mechanism to stop you poisoning or infecting yourself the same way twice (as when unlucky with prawns).

So, what mechanisms stop eating?[15] We have already met taste-specific satiety as the background for taste-specific incentive motivation. Next in line, time-wise, are receptors that detect how much your stomach is stretched, then receptors that detect how much your intestines are stretched, then receptors in the liver that signal that food absorbed into the bloodstream is being converted for use as energy. All these receptors send neural signals to the brain, directly or indirectly targeting the hypothalamus. Related hormonal signals[16] from the gut also target the hypothalamus. The hypothalamus acts as a command centre integrating all these signals. This results in a settling point that balances out all the positive and negative inputs rather than a set point based on the value of a single input like glucose.

For long-term weight control one of the most important inputs is the hormone leptin. This operates even later in the eating time-line than the stomach- and gut-related signals. Once the liver has processed your meal, any energy you don't need is stored as fat, which produces leptin. The more fat you have stored the more leptin is in your bloodstream. Leptin then acts on the hypothalamus to increase satiety signals – so, with lots of fat, eating should stop earlier. Unfortunately for my waistline, this is not a fixed fat-stat. Like all of the other eating controls, it's simply one input of many that are integrated by the hypothalamus. It is particularly unfortunate that it can be over-ridden by taste.

If you are offered food that is more palatable than your standard meal (e.g., loaded with salt, fat, and sugar), this will trigger the release of feel-good chemicals: dopamine, opiates, and serotonin (also linked to addictions). Yes, the food companies are set to addict you to their products. Tasty food will also produce resistance to satiety signals – both the shorter-term satiety signals from the gut and the longer-term leptin signal from fat. This resistance results in over-eating and, if continued, weight gain.

This weight gain can be long-term (personal data withheld). With fairly brief exposure to tasty food, rats' weights return to normal when they go back to nothing but bland boring lab chow. But given a mixed 'cafeteria' diet of similar-energy amounts of potato crisps, chocolate chip cookies, cheese crackers, *and* standard lab chow given for 90 days, relative to rats given *only* standard lab chow, they gained 20% more body weight. They kept much of this weight on when back on lab chow.[R36]

The bad news if you are thinking of dieting: with weight restriction (essentially forced dieting) applied to half the rats, they lost weight. But, back on free-feeding with lab chow, their weights increased to the same as obese rats that had not dieted (Figure 5.7). How well some of us know that yo-yo effect! The bottom line is that weight loss requires a permanent change in eating habits – and a lot of will power!

In a regular environment, then, your eating (and so weight) is the result of the interaction of habitual and metabolic signals. These signals initiate food seeking and eating (in a preparatory 'cephalic' phase driven by the brain). This hunger is then pitted against a range of taste, stomach, gut, and fat-related signals that say when you have had enough. Once feeding has started, your body enters an 'absorptive' phase, extracting nutrients and using glucose as a source of energy and storing the excess as glycogen in the liver and fat. Once all the food has been absorbed, your body enters a 'fasting' phase. Glucose is still used for energy by the brain (and made from glycogen) but the rest of the body shifts to using free fatty acids produced from the stored fat. Although there are multiple factors controlling both starting and stopping of eating, they can be seen as all operating, on different time scales, to deliver a stable body weight with adaptive quantities of fat. This can seem complicated – but check out how you feel at the various stages.

But with irregular events (like my department's celebratory morning tea) and unusually appetizing foods (particularly with the combinations of high salt, fat, and sugar beloved by junk food manufacturers), the reward system kicks in and you over-eat in metabolic terms. As with the 'cafeteria' rats, this can ultimately hijack normal control. The resultant obesity can be viewed as an addiction – an overpowering urge to activate pleasure centres,[R21] which is much closer to our view of hot complex emotion than cold simple motivation.

Hunger as an emotion[17]

You may not find it easy to think of obesity as an emotional rather than motivational disorder. But what about its opposite, anorexia (*Anorexia nervosa* = *nervous* not eating). Anorexics eat much less than they should, but not for metabolic reasons.

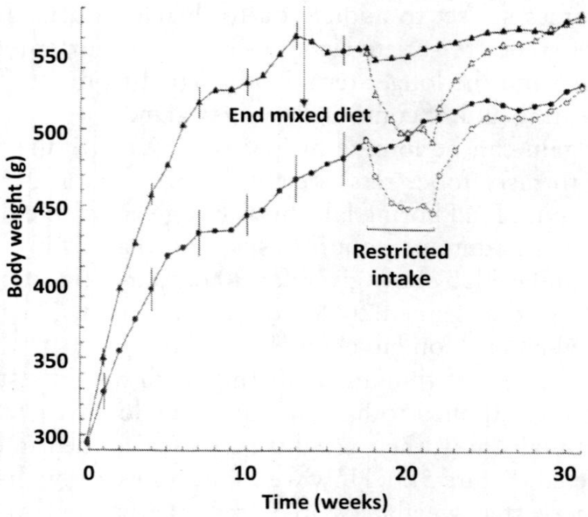

Figure 5.7 The effect of long-term 'cafeteria' food. Weight gain in rats receiving a mixed cafeteria diet (▲) or lab chow (●)for 90 days. Weight gain is maintained when shifted back to lab chow; drops when dieting and returns once dieting stops. Figure adapted from Redican[R36] with permission of the publisher via Copyright Clearance Centre.

According to the Mayo Clinic, anorexics

> have a low body weight based on personal weight history. Although many people with anorexia look very thin, some may not look thin and others may look overweight. But they've actually lost weight or failed to gain needed weight. ... If left untreated, weight loss can get to a point where [they] are at high risk of serious physical harm or death. Anorexia has the second-highest death rate of any mental illness, surpassed only by opioid overdoses. Most deaths related to anorexia stem from heart conditions and suicide. Anorexia ... can take over people's lives and be very hard to overcome.
>
> (Mayo Clinic[R27])

Anorexics do not simply ignore food. They are ambivalent: pre-occupied by food but rarely experiencing hunger. Like bulimics, they may have binge/purge cycles. They have high levels of ghrelin, but low levels of insulin, high insulin sensitivity, and feel ill after meals.

We should emphasize the emotional nature of anorexia. Anorexics are usually from groups where there is strong social pressure to be thin. They commonly have social anxiety. Many have had childhood perfectionism. They have a distorted body image. They often see themselves as fat, denying they are too thin. They have obsessive-compulsive personality patterns. Their eating is ritualized, with monotonous choices (maybe to induce taste-specific satiety), and often unusual combinations of food. Add in general anhedonia and depression. A key here may be neurotic personality,[18] positive effects of ketamine,[R34] and problems with the neurotransmitter serotonin.[R43] But the details of this are quite complicated.

Overall, then, anorexia seems to be emotional in the same sense as social anxiety disorder, obsessive-compulsive disorder, and depression – and often to have them as co-morbidities. It does not look like a simple problem with hunger motivation.[R45]

'Alimentary thinking'[19]

A final set of facts that makes me link eating with emotion are the properties of the neural systems in your gut – the enteric nervous system. This can be seen as a 'second brain' (85 billion neurons with 100 neurotransmitters) that evolved before, and may have been a spring-board for, the central nervous system (CNS). It has as many dopamine neurons as the CNS, produces 95% of the serotonin compared to 5% in your CNS, has its own blood-brain barrier, and is connected to the CNS via the vagus nerve (with 90% of the traffic being from it to the CNS).

The enteric nervous system is now seen as a key component of a Microbiome-Gut-CNS axis.[R26] This axis contributes to changes not only in hunger but also in mood and affect, particularly depression and anxiety. These changes are produced via the vagus nerve and release of ghrelin by stress. Conversely, elements of the emergency fight/flight response are generated by the enteric response to input from the CNS via the vagus. This all seems far from simple weight control.

Are emotions separate from motivations?

Emotion and motivation clearly do not mean the same thing. Given the way they have often been studied in the laboratory, it is tempting to see the words as applying to quite distinct types of state.

Emotion could be seen as reactive to external events, involving feeling and liking, and including fear, anger, happiness, and perhaps love. Motivation could be seen as homeostatic (reacting to internal imbalances of important bodily factors), involving drive and wanting, and including thirst, hunger, thermostasis, and perhaps sex (with deprivation driving behaviour to some extent but a lack not being fatal).

But as I noted for avoidance learning, fear in the emotional sense appears to be reduced when successful avoidance becomes habitual. The simplest way to see this is as habitual avoidance depending on 'fear motivation' or 'fear cognition' but without the other changes we associate with the word emotion.

That is, the full experience of emotion is linked to situations that require immediate changes in your everyday routine. Motivation is what is *left over* when you can solve the problem the same way over and over (and, with eating, at the same time every day). From this point of view, hunger can be a strong emotion (on occasion producing addiction) but is more usually habitual and so less strongly emotional.

Both fear and hunger then: (1) can be induced by external stimuli; (2) have long-term settling point like control; (3) involve autonomic activity and hormone release; (4) are less emotional and more motivational as responses become habitual; (5) involve disorders linked to serotonin. In all cases, whether fully emotional or more motivational, the emotions can be identified and their sometimes-separable parts seen as coherent, through their evolutionary function (Table 5.1).

From emotion to learning

If you view emotions as largely evolved, a question arises as to how the associated motivation can result in new responses. I talked about avoidance learning and habitual responding without mentioning how these could occur. I think this was reasonable since, clearly, a system must have evolved before its responses can be linked to stimuli by learning. But now we need to move on from emotions, with their at least initially innate reactions, to the neural plasticity that allows us to express those emotions and achieve the ends of their motivations through completely new, often arbitrary, reactions in completely new, often arbitrary, situations. That is the business of the next chapter.

Notes

1 Chapter 4, *Your triune brain*.
2 This is an additive colour code, printers use a subtractive code, see https://en.wikipedia.org/wiki/CMYK_color_model.
3 You may think of 'feelings' as being 'emotion'. However, there are feelings (like feeling cold) that are not emotional. Also, you can still experience an emotion when your feelings are blocked (though obviously it does not feel the same) – and important aspects of emotions such as anger or fear are that they involve action and specific cognitions about sources and appropriate actions towards those sources.
4 *Separating the Parts – Separation Anxiety*.
5 CORT here means corticosterone; in humans the equivalent hormone is cortisol. Calling them both CORT should make the parallel arguments clearer.
6 I would be happy to call these cognitions and beliefs, but for most people this would be going too far.
7 For a detailed table of different types of aggression and defensive behaviour and their functions, see Table 17.2 in Pinel.[R33]
8 This is the term preferred, for simplicity, by Redican[R35, p. 239] rather than the more formal 'silent bared-teeth face'.
9 As put by the authors of Figure 5.2 "In contrast with suggestions that humans are endowed with an intricate facial musculature for the purposes of social communication, here we examined the idea that emotional expressions not only signal emotions, but may have originated to modify preparedness for perception

and action. Specifically, we sought evidence for Darwin's principles of expression opposition in form and function. Fear and disgust were shown to be near opposites in form, supported by fundamentally opposing action tendencies. Evidence of enhanced visual-field size, saccadic velocity and nasal inspiratory capacity in fear and the direct inverse in disgust indicate that there is a parallel opposition in function between the two expressions, serving the functions of sensory vigilance to enhance detection of the source of potential threat and sensory rejection to reduce sensory exposure".[R42, p. 847]

10 At my age you are likely to meet a pharmacological equivalent of this balancing action. You can take the anticoagulant dabigatran to prevent clots and stroke. But if you do something that causes severe bleeding, or if you are about to have surgery, its effects can be reversed by giving an antagonist. Most other anticoagulants don't have an antagonist but may be better if you have kidney disease or a prosthetic heart valve.

11 In Bard's own writing, the theory itself is attributed to Cannon. But it is elaborated in most detail by Bard.[R1]

12 For example, https://www.ncbi.nlm.nih.gov/books/NBK519712/table/ch3.t11/.

13 For a recent review of insulin as "The master regulator of glucose metabolism", see Norton et al.[R31]

14 "These include the set-point, dynamic equilibrium, adiposity force, control theory-settling point, Hall-Guo, operation point and dual intervention point (DIP) models. The set-point model posits a single reference point around which levels of fat are regulated. The dynamic equilibrium model suggests that the apparent regulation of body fat around a reference point is an illusion owing to the necessary impacts of weight change on energy expenditure. Control theory focuses on the importance of feedback gain and suggests set-point and dynamic equilibrium are ends of a continuum of feedback gain. Control theory models have also been called 'settling point' models. The Hall-Guo, operation point and DIP models also bring together the set-point and dynamic equilibrium ideas into a single framework. The DIP proposes a zone of indifference where dynamic equilibrium 'regulation' predominates, bounded by upper and lower intervention points beyond which physiological mechanisms are activated."[R41, p. 1]

15 For more details of the content of the next two paragraphs, and for some nice coloured diagrams, see Flier and Maratos-Flier.[R15]

16 Cholecystokinin (CCK), peptide YY (PYY), glucose.

17 The bulk of the content of this section is based on overview paragraphs from Kaye et al.[R20] The main focus of the paper is the neurobiology of anorexia and, particularly, the serotonin system. The overview paragraphs provide sources for more detail on the other aspects of anorexia that I summarize here.

18 See Chapter 10, *Neuroticism – A key personality factor for internalizing disorders*.

19 This section is based largely on https://www.newscientist.com/article/mg21628951-900-gut-instincts-the-secrets-of-your-second-brain/, which has a nice coloured figure comparing the enteric and central nervous systems.

References

R1. Bard, P, Emotion, I, The neuro-humoral basis of emotional reactions, in *A handbook of general experimental psychology*. 1934, Worcester: Clark University Press. p. 264–311.

R2. Bhutani, S, Wells, N, Finlayson, G, & Schoeller, DA, Change in eating pattern as a contributor to energy intake and weight gain during the winter holiday period in obese adults. *International Journal of Obesity*, 2020. **44**: 1586–95.

R3. Brady, JV, Conditioning and emotion, in *Emotions: Their parameters and measurement*, L Levi, Editor. 1975, New York: Raven Press. p. 309–40.
R4. Brady, JV, Toward a behavioural biology of emotion, in *Emotions: Their parameters and measurement*, L Levi, Editor. 1975, New York: Raven Press. p. 17–46.
R5. Cannon, WB, The James-Lange theory of emotion: A critical examination and alternative theory. *American Journal of Psychology*, 1927. **39**: 106–24.
R6. Cannon, WB, *Bodily changes in pain, hunger, fear and rage*. 1936, New York: Appleton-Century.
R7. Chevalier-Skolnikoff, S, Facial expression of emotion in non-human primates, in *Darwin and facial expression*, P Ekman, Editor. 1973, New York: Academic Press. p. 11–89.
R8. Coover, GD, Goldman, L, & Levine, S, Plasma corticosterone increases produced by extinction of operant behaviour in rats. *Physiology and Behavior*, 1971. **6**: 261–3.
R9. Coover, GD, Ursin, H, & Levine, S, Plasma corticosterone levels during active avoidance learning in rats. *Journal of Comparative and Physiological Psychology*, 1973. **82**: 170–4.
R10. Darwin, C, *The expression of the emotions in man and animals*. 1965/1872, Chicago: University of Chicago Press.
R11. Davitz, JR, *The language of emotion*. 1969, New York: Academic Press.
R12. Dawkins, R, *The blind watchmaker*. 1986, Harlow: Longman Scientific & Technical.
R13. Eibl-Eibesfeldt, I, *Love and hate: The natural history of behavior patterns*. 1997, London: Methuen &Co Ltd.
R14. Ekman, P & Friesen, WV, A new pan-cultural facial expression of emotion. *Motivation and Emotion*, 1986. **10**: 159–68.
R15. Flier, JS & Maratos-Flier, E, What fuels fat. *Scientific American*, 2007. **297**: 72–81.
R16. Hofer, MA, Physiological and behavioural processes in early maternal deprivation, in *Physiology, emotion and psychosomatic illness. Ciba symposium No. 8 (new series)*, R Porter & J Knight, Editors. 1972, Amsterdam: Elsevier. p. 175–86.
R17. Hohmann, GW, Some effects of spinal cord lesions on experienced emotional feelings. *Psychophysiology*, 1966. **3**: 143–56.
R18. James, W, What is an emotion? *Mind* (reprinted in Arnold, 1968, The nature of emotion), 1884. **9**: 188–205.
R19. Kawai, N, The underlying neuronal circuits of fear learning and the Snake Detection Theory (SDT), in *The fear of snakes*, N Kawai, Editor. 2019, Singapore: Springer Nature Singapore Pte Ltd. p. 33–58.
R20. Kaye, WH, Frank, GK, Bailer, UF, & Henry, SE, Neurobiology of anorexia nervosa: Clinical implications of alterations of the function of serotonin and other neuronal systems. *International Journal of Eating Disorders*, 2005. **37**: S15–9.
R21. Kenny, PJ, The food addiction. *Scientific American*, 2013. **309**: 34–9.
R22. Kret, ME, Prochazkova, E, Sterck, EHM, & Clay, Z, Emotional expressions in human and non-human great apes. *Neuroscience & Biobehavioral Reviews*, 2020. **115**: 378–95.
R23. Lader, M & Tyrer, P, Vegetative system and emotion, in *Emotions: Their parameters and measurement*, L Levi, Editor. 1975, New York: Raven Press.
R24. LeDoux, JE, Rethinking the emotional brain. *Neuron*, 2012. **73**: 653–76.

R25. Mandler, G, *Mind and body*. 1984, New York: Norton.
R26. Margolis, KG, Cryan, JF, & Mayer, EA, The microbiota-gut-brain axis: From motility to mood. *Gastroenterology*, 2021. **160**: 1486–501.
R27. Mayo Clinic. *Anorexia nervosa*. [cited 2025]; Available from: https://www.mayoclinic.org/diseases-conditions/anorexia-nervosa/symptoms-causes/syc-20353591
R28. McNaughton, N, *Biology and emotion*. 1989, Cambridge: Cambridge University Press.
R29. McNaughton, N, What do you mean "anxiety"? Developing the first anxiety syndrome biomarker. *Journal of the Royal Society of New Zealand*, 2018. **48**: 177–90.
R30. McNaughton, N, Personality neuroscience and psychopathology: Should we start with biology and look for neural-level factors? *Personality Neuroscience*, 2020. **3**: e4.
R31. Norton, L, Shannon, C, Gastaldelli, A, & DeFronzo, RA, Insulin: The master regulator of glucose metabolism. *Metabolism*, 2022. **129**: 155142.
R32. Pinel, JPJ, *Biopsychology* (7th ed., 8th ed.). 2007, Boston: Allyn and Bacon
R33. Pinel, JPJ, *Biopsychology* (8th ed.). 2011, Boston: Allyn and Bacon (Pearson).
R34. Ragnhildstveit, A, Slayton, M, Jackson, LK, Brendle, M, Ahuja, S, Holle, W, Moore, C, Sollars, K, Seli, P, & Robison, R, Ketamine as a novel psychopharmacotherapy for eating disorders: Evidence and future directions. *Brain Sciences*, 2022. **12**: e382.
R35. Redican, WK, An evolutionary perspective on human facial displays, in *Emotion in the human face*, P Ekman, Editor. 1982, Cambridge: Cambridge University Press. p. 212–280.
R36. Rolls, BJ, Rowe, EA, & Turner, RC, Persistent obesity in rats following a period of consumption of a mixed, high energy diet. *The Journal of Physiology*, 1980. **298**: 415–27.
R37. Rolls, ET, Emotion, motivation, decision-making, the orbitofrontal cortex, anterior cingulate cortex, and the amygdala. *Brain Structure and Function*, 2023. **228**: 1201–57.
R38. Rutledge, LL & Hupka, RB, The facial feedback hypothesis: Methodological concerns and new supporting evidence. *Motivation and Emotion*, 1985. **9**: 219–40.
R39. Secombe, P, Woodman, R, Chan, S, Pilcher, D, & van Haren, F, *Epidemiology and outcomes of obese critically ill patients in Australia and New Zealand. Critical Care and Resuscitation*, 2020. **22**: 35–44.
R40. Skvortsova, A, Veldhuijzen, DS, Kloosterman, IEM, Pacheco-López, G, & Evers, AWM, Food anticipatory hormonal responses: A systematic review of animal and human studies. *Neuroscience & Biobehavioral Reviews*, 2021. **126**: 447–64.
R41. Speakman, JR & Hall, KD, Models of body weight and fatness regulation. *Philosophical Transactions of the Royal Society B: Biological Sciences*, 2023. **378**: 20220231.
R42. Susskind, JM, Lee, DH, Cusi, A, Feiman, R, Grabski, W, & Anderson, AK, Expressing fear enhances sensory acquisition. *Nature Neuroscience*, 2008. **11**: 843–50.
R43. Underwood, MD, Kassir, SA, Bakalian, MJ, Galfalvy, H, Dwork, AJ, Mann, JJ, & Arango, V, Serotonin receptors and suicide, major depression, alcohol use

disorder and reported early life adversity. *Translational Psychiatry*, 2018. 8: e279.

R44. Weingarten, HP & Martin, GM, Mechanisms of conditioned meal initiation. *Physiology & Behavior*, 1989. 45: 735–40.

R45. Zhang, Z, Robinson, L, Campbell, I, Irish, M, Bobou, M, Winterer, J, Zhang, Y, King, S, Vaidya, N, Broulidakis, MJ, van Noort, BM, Stringaris, A, Banaschewski, T, Bokde, ALW, Brühl, R, Fröhner, JH, Grigis, A, Garavan, H, Gowland, P, Heinz, A, Hohmann, S, Martinot, J-L, Martinot, M-LP, Nees, F, Orfanos, DP, Paus, T, Poustka, L, Sinclair, J, Smolka, MN, Walter, H, Whelan, R, Schumann, G, Schmidt, U, & Desrivières, S, Distinct personality profiles associated with disease risk and diagnostic status in eating disorders. *Journal of Affective Disorders*, 2024. 360: 146–55.

6 How do we learn?

*This chapter is very technical. It explains mental processes at the molecular level. You may want to read the first section and then skip to "Three rules – several learning patterns". Particularly technical sections have their headings in italics with a *. But if you really want to understand mental mechanics, this is for you.*

From emotion to learning

Evolution can, as you saw in Chapter 5, account for our specific fear of snakes (or a rat's fear of a cat) – or why we easily develop a specific fear of spiders. Less easily, we learn to fear recent dangers (e.g., guns) that our ancestors did not have to face and so are not evolved fear reactions. The same is true for many stimuli that can elicit your other emotions. They are evolutionarily novel and must be learned. But that learning taps into highly evolved, emotion-specific, action systems. These mental mechanics can help us understand emotion as well as the broad range of stimuli that can elicit it.

> Despite millennia of preoccupation with every facet of human emotion, we are still far from explaining in a rigorous physiological sense this part of our mental experience. Neuroscientists have, in modern times, been especially concerned with the neural basis of cognitive processes such as perception and memory. They have for the most part ignored the brain's role in emotion.
>
> Yet in recent years, interest in this mysterious mental terrain has surged. Catalyzed by breakthroughs in understanding the neural basis of cognition and by an increasingly sophisticated knowledge of the anatomical organization and physiology of the brain, investigators have begun to tackle the problem of emotion. One quite rewarding area of research has been the inquiry into the relation between memory and emotion. Much of this examination has involved studies of one particular emotion -- fear -- and the manner in which specific events or stimuli come, through individual learning experiences to evoke this state.
>
> (LeDoux[R13, p. 32])

DOI: 10.4324/9781003465027-6

Table 6.1 Learning as simplification. A comparison of the complexity of human fear-related disorder (SANDRA), naturally occurring unlearned fears (INNATE), and laboratory-conditioned fear (CONDITIONED)

	Sandra	*Innate*	*Conditioned*
Nature of stimuli	Unknown	Complex	Simple/controlled
Timing of stimuli	Unknown	Un/controlled	Controlled
Skeletal responses	Complex	Uncontrolled	Controlled

The study of fear learning in rats simplifies both the stimuli and threats involved. In the laboratory, 'conditioned fear' can be tightly controlled (Table 6.1).

Adaptive innate reactions are more complex. Innate fear reactions are technically difficult to study.[1] The cat that elicits fear in a rat has a mind of its own. Its behaviour and the consequences of its action depend both on the situation and the behaviour of the rat.

Maladaptive human clinical cases are very complex. Sandra Edwards, with the backing of her psychiatrist, published her 'sufferer's perspective' of panic disorder and agoraphobia in Journal of Affective Disorders.[R5] It shows the complexity of a typical clinical case. It[2] takes nine journal pages to vividly describe "the clinical features, the heritable components, the environmental contributions, the developmental penalties, the social consequences, and the therapies for these conditions".[R8, p. 65]

So, let's look at fear conditioning in detail. This will provide us with three basic learning rules in terms of mental mechanics. These explain a wide range of learning paradigms. If you just want to understand the basic rules and apply them to the paradigms, skip to the diagrams in Figure 6.7. If you want to understand how the rules come about, read on.

Classical fear conditioning – from Pavlov to brain circuits

In classical fear conditioning, you deliver an essentially neutral stimulus (e.g., a tone) to signal the upcoming delivery of a brief painful stimulus (usually a shock). This is like Pavlov's bell-food pairing, where, after training, the bell made his dogs salivate even if food was no longer delivered. Classical conditioning is sometimes called Pavlovian conditioning.

The most basic classical conditioning involves 'stimulus substitution'. The previously neutral stimulus (now a conditioned stimulus; CS+) produces much the same conditioned responses (CRs) as the shock (unconditioned stimulus, US) did. The tone increases blood pressure and decreases movement about the experimental chamber. This CS+/CR relationship is different from instrumental conditioning (which I will talk about later), where completely new responses (like pressing a lever) can be learned. It is associative in the sense that if the tone were presented unpaired with the shock (making it a CS–), it would not acquire the CRs.

Joe Ledoux's laboratory searched for the neural circuits involved in tone-shock conditioning (for an overview see LeDoux).[R13] Tones are detected in the ear and the information sent to the midbrain and then auditory thalamus (like the snake detection circuit[3]). Lesion of either of these blocked the tone conditioning. However, lesions of the auditory cortex did not. As with the snake detection circuit, the thalamus sends information not only to the cortex but also directly to the amygdala. Lesions of the amygdala appeared to block fear conditioning but they also blocked unconditioned fear. The amygdala looked to be where the tone got connected to the effects of the shock.

Classical fear conditioning – neural plasticity

The thalamus-amygdala pathway also looks as though it can 'learn'. It shows long-term potentiation (LTP). LTP is a change in synaptic strength that lasts for a long time (hours, days, and sometimes a year or so). It was discovered by accident in experiments stimulating a single pathway to produce frequency potentiation.[4] It was later shown to be associative in the sense required for Joe Ledoux's example of fear conditioning. At the synaptic level, LTP is very simple. It could occur at just a single synapse.

Blocking a specific type of glutamate receptor, the NMDA[5] receptor, blocks LTP. Importantly, it also blocks fear conditioning. So, here, LTP is the glue that binds the tone to the memory of shock. If we look closely at the mechanisms of LTP, we can see the NMDA receptor as a simple coincidence detector. So, classical conditioning, in this particular case, is the result of pure association. This is mental mechanics at the molecular level.

In what follows, I will be contrasting NMDA receptors with AMPA[5] receptors. Both of these types of receptor work with the excitatory neurotransmitter glutamate. (Yes, that is the same compound that, with added sodium, is the MSG you likely have in your kitchen cupboard.) Just knowing that there are two different types of glutamate receptor and that the NMDA one can be blocked selectively is all you really need to know to understand drug manipulations of LTP.

If you don't want to know the molecular details of how LTP works, just skip to the next section. But, if you do want to know the details, or how neurotransmitter receptors work in general you need to know how receptor types work and are named.

**Receptor types (some general pharmacology)*[6]

Transmitter receptors are complex molecules that evolved a shape to bind their particular transmitter. But these molecules can be slightly different when they arise (at different times) in different parts of the brain. This is great news for pharmaceutical companies as they can design drugs that bind to just one such receptor type and produce a more selective action (Figure 6.1).

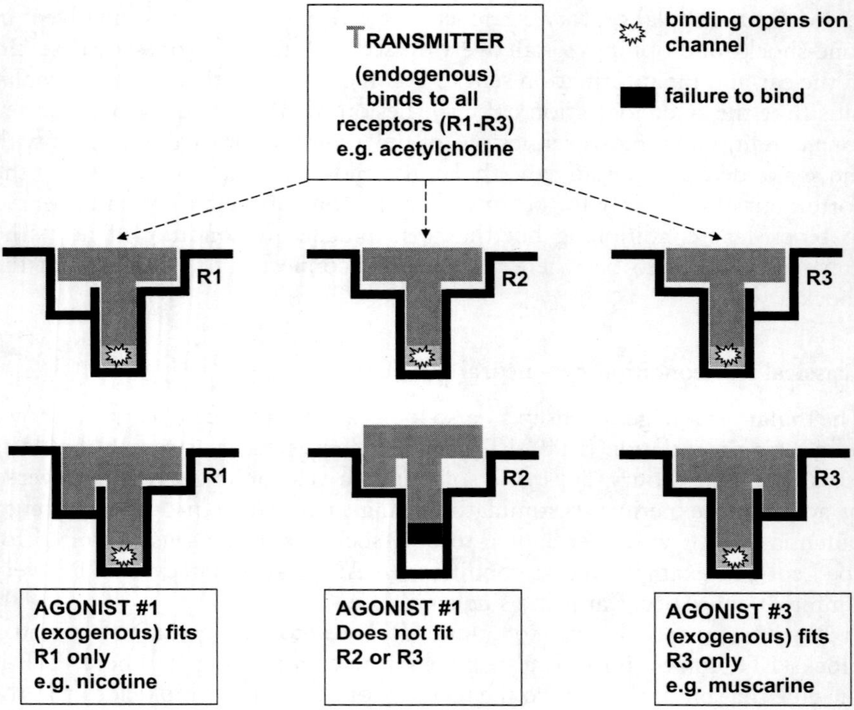

Figure 6.1 Receptor subtypes. Transmitters, receptors, and exogenous drugs have effects that depend on the shape of their molecules. The transmitter acts like a master key in all the receptor locks. Pharmaceutical compounds are designed to be like single-door keys. Binding causes the opening of channels, with excitation and action potentials dependent on sodium channels (see Figure 3.2).

This also accounts for the selective actions of some naturally occurring drugs, like aspirin, that get discovered rather than synthesized. Often, traditional medicine flags such compounds and their uses.

Receptors are often named for the compound that was first discovered to bind to them. Nicotine (from the tobacco plant) binds to a receptor of the transmitter acetylcholine. So, this is called the nicotinic acetylcholine receptor (or just nicotinic receptor, for short). The receptor just happens to bind nicotine. It evolved independently of our use of tobacco! Similarly, muscarine named the muscarinic receptor for acetylcholine. Muscarine comes from certain mushrooms (some deadly), particularly the usually red-capped *Amanita muscaria*. Unlike nicotine, it causes hallucinations and seizures. Again, the evolution of muscarine and of the receptors was completely independent.

Blocking the muscarinic receptor can also have nasty effects. The deadly nightshade family are toxic because they produce a muscarinic antagonist, atropine. But atropine is not all bad – in low doses. During the renaissance women used an extract from deadly nightshade to enlarge their pupils for cosmetic reasons. One idea is that, since sexual attraction dilates the pupils, this eye enlargement will purposefully send a false signal that you are attracted to the man and this makes you attractive to male egos. The Latin name of deadly nightshade is *Atropa belladonna* (bella donna = beautiful lady). Atropine also treats poisoning (by, e.g., the muscarine in mushrooms), unstable heart slowing, and excess bodily secretions of various types. It is used to paralyse the eye's accommodation reflex, and to dilate the pupils for non-cosmetic reasons.

Many receptors are not named but are just given letters (e.g., α, β – as in the beta blockers people are given for blood pressure problems), or numbers, or both, added after the transmitter name. So, you don't have to remember which transmitter they belong to. For example, mammalian serotonin/5HT receptors are labelled as families from $5HT_1$ to $5HT_7$ with, for example, the $5HT_1$ family ranging from $5HT_{1A}$ to $5HT_{1F}$ receptors.

The only thing you really need to remember about these receptors is that there are a lot of them and so a lot of possibilities for drug companies to play with. But, with receptors named after chemicals that bind specifically to them, you need to remember the transmitter – glutamate for both NMDA and AMPA receptors (with NMDA and AMPA being artificial chemicals that are selective for their type of receptor) and acetylcholine for both nicotinic and muscarinic receptors.

**How does long-term potentiation work?*

In molecular terms, LTP requires both AMPA and NMDA receptors. Glutamate binding to the post-synaptic AMPA receptor produces excitation and, to the NMDA receptor, produces an influx of calcium (Ca++) – but *only if* the post-synaptic cell is sufficiently depolarized via AMPA receptors. So, the NMDA receptor acts as a coincidence detector. If you don't want to know precisely how it does this, feel free to skip to the next section.

The AMPA receptor is linked to a sodium channel (Figure 6.2A, B) and so produces post-synaptic depolarization.[7] The NMDA receptor is linked to a calcium channel (which allows Ca++ through but not Na+). A crucial feature of the Ca++ channel is that it is normally blocked by a magnesium ion (Mg++), like a cork in a bottle (Figure 6.2C). So, when glutamate binds to the NMDA receptor, the open channel does not let Ca++ through. In contrast, the open AMPA receptor always lets Na+ through.

When the post-synaptic cell fires an action potential, this produces a large positive voltage in the post-synaptic cell. This repels the positive Mg++ ion (as happens when you put the positive ends of two magnets together) and allows Ca++ to flow into the cell (Figure 6.2D). This triggers a complicated biochemical cascade that eventually feeds back to the pre-synaptic terminal (Figure 6.2E), ultimately making it stronger (Figure 6.2F). That is, when an action potential in the pre-synaptic cell causes the terminal to release

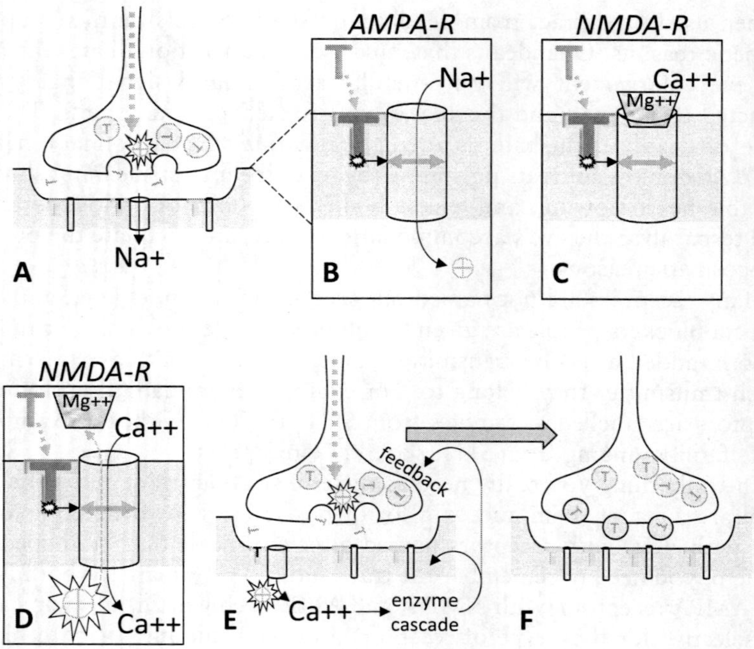

Figure 6.2 Long-term potentiation. (A) Diagrammatic cross-section of an individual synapse showing arrival of action potential, binding of transmitter to post-synaptic AMPA receptor (AMPA-R), and passage of sodium (Na+) through the open ion channel. (B) Magnified diagram of open sodium channel. (C) Equivalent with binding to post-synaptic NMDA receptor (NMDA-R). Passage of calcium (Ca++) through the channel is blocked by magnesium (Mg++) 'cork'. (D) As C but with post-synaptic depolarization by action potential, which blows out the 'cork' – Ca++ now passes through the channel. (E) Post-synaptic Ca++ signal triggers an enzyme cascade producing post- and pre-synaptic feedback; and (F) resulting in strengthening of the synapse. Symbols: T = neurotransmitter (glutamate); **T** = receptor for transmitter; dashed grey arrow = arrival of pre-synaptic action potential; starburst = action potential; ⊕ = positive charge; cylinder = ion channel; ⇔ = channel opened by T.

neurotransmitter *at the same time* as the post-synaptic cell has an action potential, the connection between the two is strengthened.

From LTP to association

Knowing the molecular mechanism allows us to see that the NMDA-receptor-Ca++ channel complex is a coincidence detector. Importantly, when it detects coincident pre- and post-synaptic activity (coding tone and shock reaction, respectively, in Ledoux's experiment), it 'rewires' the synapse. This produces an association between the two stimuli. That is, after this strengthening, the pre-synaptic input (e.g., tone) can activate the post-synaptic cell by itself and so produce response components like those of the shock.

This is a physical example of what is termed a Hebb synapse – as hypothesized on purely theoretical grounds in 1949 by Donald Hebb.[8]

> The general idea is an old one[9], that any two cells or systems of cells that are repeatedly active at the same time will tend to become 'associated'. ... [More formally,] When an axon of cell A is near enough to excite cell B and repeatedly or persistently takes part in firing it, some growth process or metabolic change takes place in one or both cells such that A's efficiency, as one of the cells firing B, is increased ... [It is] most probable ... that synaptic knobs develop and increase the area of contact between the [cells].
>
> (Hebb[R10, pp. 62, 70])

Carla Shatz coined a neat phrase (sometimes wrongly attributed to Hebb) that captures this process: "Cells that fire together wire together". That is, "the timing of action potential activity is critical in determining which synaptic connections are strengthened and retained and which are weakened and eliminated"[R18, p.64] Hence the mirror phrase "cells out of synch lose their link". These two phrases capture both the simple associative learning in Joe Ledoux's experiments and the associative experience-dependent visual development that Shatz was studying, but the detailed rules of the two are not identical.

An important point about a simple associative mechanism for classical conditioning is that it must produce 'stimulus substitution' of the type originally proposed by Pavlov.[10] The previously neutral stimulus takes the place of the unconditioned stimulus and so must produce the same (or some of the same) responses. It does not involve strengthening of connections by a 'reinforcer'. This can also occur and we will come to that in the next section.

Stimulus substitution and 'sensory pre-conditioning'

Stimulus substitution is one explanation for a stimulus acquiring a *conditioned response*. But Hebb's original example of association, and Shatz's developmental work (covered below as 'perceptual learning'), focused on the coincidence of pairs of visual stimuli or just correlated spontaneous activity ("any two cells or systems of cells"). Association should not need to have a significant event like a shock paired with a visual or auditory stimulus. So, it should not need to produce a change in responding. The substitution simply provides behavioural evidence that the association occurred.

When you present a rat with a regular pairing of a tone and a light, its behaviour shows no obvious changes. So, a radical behaviourist would say there had been no learning. However, if you experienced the same pairing, you would come to expect the light after the tone. We can check this by asking you about it. But, how do you ask the rat the same question? How do you find out what is in its mind? Here is an example.[R16] The result has been called 'sensory pre-conditioning'[R4] – but I personally see it as just as much tone-light conditioning as Ledoux's is tone-shock conditioning.

> *Phase 1:* Stimulus A (a tone or light) is paired with stimulus B (a light or tone, respectively) in a series of classical (Pavlovian) conditioning-like trials. Neither A nor B produces any observable response, before or after the conditioning-like trials. *Conclusion: You cannot tell if conditioning has occurred or not.*
>
> *Phase 2:* Stimulus B is next paired with shock in a series of conditioning trials while the thirsty rat is licking a spout for water. After a number of trials, B suppresses the rat's licking. *Conclusion: Conditioning has occurred.*
>
> *Phase 3:* Stimulus A is now presented to the subject without any previous pairing of A with shock. A suppresses the rat's licking although A has never been paired with shock. *Conclusion: Conditioning must have occurred through simple association in phase 1, but was behaviourally silent.*

The changes in neuronal connections that we can assume have occurred are shown in Figure 6.3.

So, simple association resulting from synaptic strengthening produces long-term changes in the 'knowledge' encoded by the brain. Association will only show up as changed behaviour if there is a reason for the behaviour. You can be asked later in language if you noticed the pairing of events. As we have seen, a rat can also be asked later via an appropriate presentation of significant events. What I have just called significant events are also usually termed 'reinforcers' and learning is often seen as requiring reinforcement.

It turns out that classical conditioning *can* involve reinforcement rather than simple association – so let's look at the mental mechanics of Activity Dependent Facilitation (ADF) for that. ADF is like LTP in being a long-term associative increase in the strength of a pre-synaptic terminal. However, the association is driven by the firing not of a post-synaptic cell but of a 'facilitator' neuron. The facilitator is activated by input from the US (the shock), and then releases 5HT onto the pre-synaptic terminal. The 5HT acts as a reinforcer of the connection between the pre- and post-synaptic cells. As a result, the post-synaptic cell (controlling the CR) does not have to fire[R6] and so the CR itself does not need to match the UR and so can avoid stimulus substitution. This has been studied at the single neuron level in clever experiments, particularly by Eric Kandel for an overview see Carew et al.,[R12] taking advantage of the simplicity of the nervous system of a sea slug *Aplysia californica*. If this seems too detailed for you, just skip to the next section.

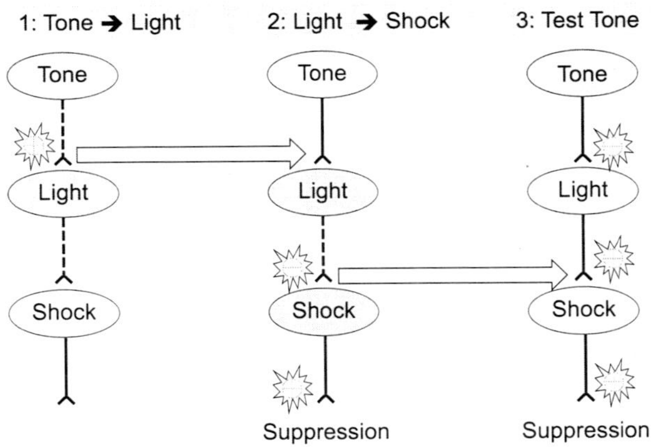

Figure 6.3 Sensory pre-conditioning. (1) Phase 1: Light and tone are paired (with light preceding tone). This strengthens their latent connection (dashed line). (2) Phase 2: Shock produces suppression and preceding it with the tone strengthens their latent connection (dashed line). Light is not activated and so its previously strengthened connection is not active. (3) Phase 3: Light is presented without tone or shock. It activates the tone representation which activates the shock representation which produces suppression. The same kind of results are obtained if food is presented not shock and salivation measured (as in Pavlov's experiments) not suppression. ⊕ = synapse activated. Filled ovals = external stimulus input.

**From Activity Dependent Facilitation to classical reinforcement*

In *Aplysia* some of the single neurons (particularly the facilitator neurons) can be reliably identified across different *Aplysia*. This allows carefully repeated experiments starting with 'the same' fresh neuron each time. The entire circuit involved in classical conditioning has been identified (Figure 6.4).

The associative nature of the conditioning can be shown by using a light touch to either the siphon or the mantle as the CS+ and then testing both after tail shock conditioning. Only the one used as the CS+ shows the CR and not the one used as a CS- despite both receiving 5HT. Swapping the two (in a new *Aplysia*) shows that it is the pairing with shock that is important, not the place touched.

From classical to instrumental reinforcement

Care is needed here. At the behavioural level, both LTP and ADF can support an association that results in a CS+ producing a CR. If, in the ADF case, the UR to the shock is the same as the CR, it will be difficult to tell the result apart from that produced by LTP. It is only mental mechanics at the synaptic level that allows us to be sure that Ledoux's tone-shock conditioning is *simple association*; while Kandel's touch-shock conditioning is *reinforcement*. LTP

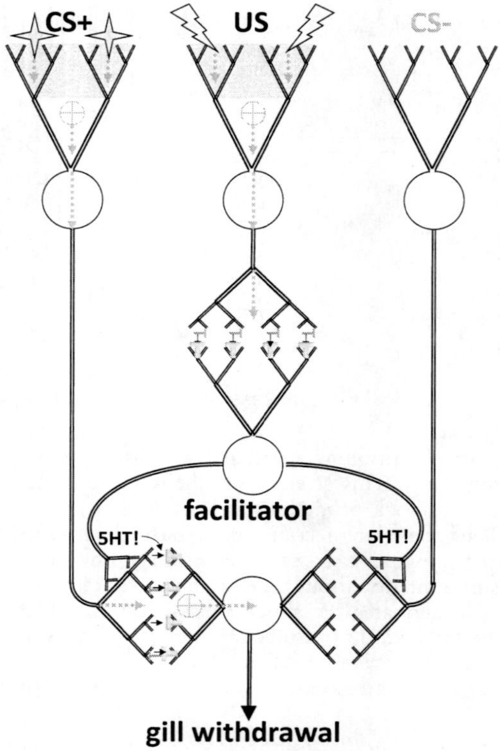

Figure 6.4 Classical conditioning in Aplysia.[R13] A tail shock (US ⚡) fires an interneuron (facilitator), releases 5HT onto the terminals of the mantle and siphon. If the mantle is touched just before, making it a CS+ (✦), the 5HT strengthens its connections to the neuron that controls gill withdrawal. The CS+ then produces withdrawal on future occasions. The 5HT does not affect the CS– because they are not paired. You can make the siphon the CS+ and mantle the CS- to show the stimulus-specificity of the association.

accounts for Pavlov's explanation of stimulus substitution. ADF accounts for cases where Pavlov's rule fails. ADF does not fit Hebb's rule.[R6]

Whether the CR is like the UR or not, the experimenter has no control over the response. The response is determined by the US. (Food produces salivation; shock does not.) This aspect of classical conditioning contrasts with instrumental conditioning – where the experimenter (or circus trainer or sheep dog handler) picks a response that they want to control and pairs it with presentation of a reward.[11] That is, the CR is followed by the US, rather than the other way round. To make this happen, the original response needs to occur spontaneously, without the US. It often occurs driven by stimuli in the training situation. Following it with a reward then increases its frequency. Often, the trainer only gives a reward after a specific signal, so that they can get the trainee to make the response on demand.

If you want to train a new response that is not normally emitted, you need to do what is called shaping. In its simplest and oldest form, you give a reward whenever what the animal does is in the direction that you want and then give the reward only as it gets closer and closer to what you want it to do – slowly 'shaping' its behaviour to the right pattern.

With rats pressing levers, it's common to 'cheat' with this and do things like smear a bit of tasty food on the lever at the very start of training. When the rat tries to eat this, it often presses the lever and then you (or nowadays your computer) deliver the reward. We also now cheat with what is called autoshaping – see later.

A model system for instrumental reinforcement

To show how instrumental conditioning works, the discovery of intracranial self-stimulation (ICSS) was crucial. Electrical stimulation of key brain areas is rewarding. Both rats and humans will happily stimulate themselves repeatedly, potentially to exhaustion. You can shape rats to press a lever to do this (Figure 6.5A; for videos of this, search for 'intracranial self-stimulation' on YouTube).

ICSS has been used to track down the mechanism of instrumental reinforcement in the brain.[R17] John Reynolds, Brian Hyland, and Jeff Wickens[12] selected rats with strongly reinforcing ICSS placements in the substantia nigra and, under anaesthesia, recorded from inside neurons that had input from cortical neurones and which then sent output to the striatum to generate responding. Importantly, the nigral input to the striatum releases dopamine onto the *synapses* of the cortical cells (Figure 6.5B). This pre-synaptic dopamine release is similar to the serotonin system in *Aplysia* and so does not by itself produce strengthening of the cortical input. This is just as well as dopamine is released in many brain areas at once and not targeted to the synapses we want strengthened.

Innate stimuli (or pre-learned ones, or in this experiment current injection) must depolarize the striatal output cell before ICSS can strengthen the previously neutral CS+ input (cortical stimulation in this experiment), provided it is also active. To make sure that the ICSS effect was produced by dopamine release, in this experiment, they blocked its effect with the specific dopamine antagonist SCH23390[13] (Figure 6.5C).

**Combining ADF with LTP for instrumental reinforcement*

So, we have the combination of pre- and post-synaptic depolarization that often produces LTP, plus the pre-synaptic neuromodulation that often produces ADF. To make this clearer, let's look at an experiment that simply pulsed a small amount of dopamine onto slices of cortex plus striatum circuitry, while stimulating the cortical input and recording from the striatum.[R19] Single pulses to the cortical input produced a stable response

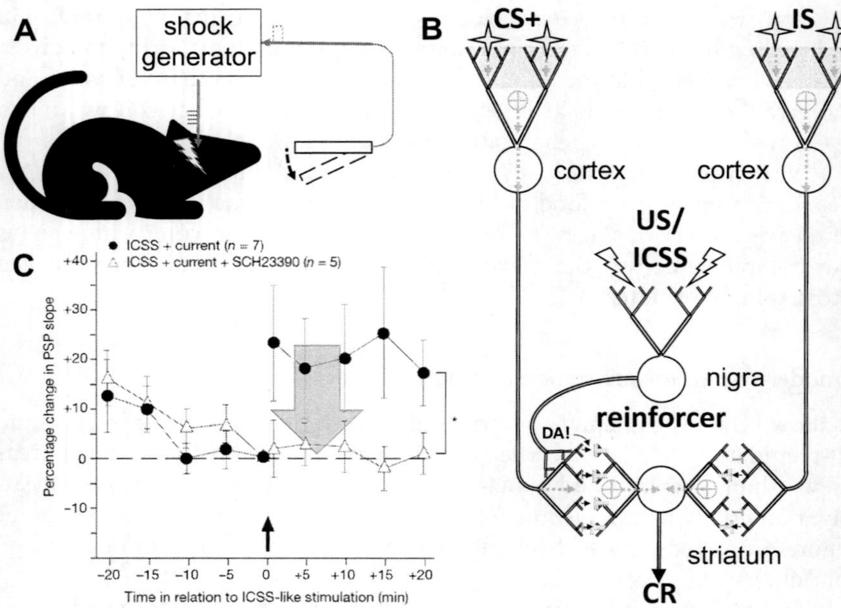

Figure 6.5 Mechanisms of Instrumental Reinforcement. (A) Intracranial self-stimulation (ICSS). If a rat presses the lever (downward arrow), then an electrical pulse (dashed rectangle) is sent to a shock generator and triggers a burst of pulses (||||) that are delivered as a shock (↯) to a reward centre in the rat's brain. The rat will then keep pressing the lever to get the rewarding feeling (humans will do the same). (B) Circuits involved in ICSS and rewarded learning. Innate stimuli (IS; or pre-learned ones) can generate a response via the striatum. If this response is followed by ICSS to the substantia nigra (or by a natural reward, US), this releases dopamine onto terminals of cortical neurons that project to the striatal cells. Provided a stimulus (CS+) has activated the cortical input, its connection to the response is strengthened and, in future, presenting the CS+ without the IS can generate a conditioned response (CR). (C) Cortical stimulation by itself does not produce an increased response over time (min. 20 to 0), but if it is paired with one presentation of ICSS-like stimulation (🡅) it then shows a long-term increase in response (min 0 to +20). However, if a dopamine-blocking drug is given beforehand this increase is blocked (🡇) although the original cortical response is unchanged. Graph from Jenkins and Moore[R17] used with permission of Springer Nature BV; permission conveyed through Copyright Clearance Center, Inc.

(Figure 6.6, min −10 to 0) and then a high-frequency train of pulses *coupled with post-synaptic depolarization* was applied at minute 0. This had no immediate effect but between 5 and 20 minutes produced depression of the response (i.e., LTD). The same protocol in hippocampal slices would likely have produced LTP. If a small drop of potassium solution was applied at the same time, there was a slight initial increase but, by

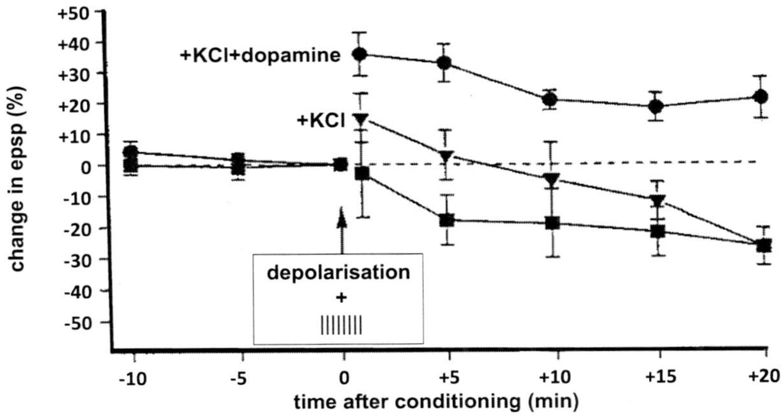

Figure 6.6 Pre-synaptic dopamine plus pre- and post-synaptic depolarization. −10 to 0 minutes: single pulse stimulation of cortical input to striatum. 0 minutes: conditioning with high-frequency stimulation (|||||) + depolarization, alone (■), with a pulse of potassium chloride (KCl, ▲) or with KCl and dopamine (●). Figure adapted from Wickens et al.[R19] used with permission of Elsevier Science & Technology Journals; permission conveyed through Copyright Clearance Center, Inc.

20 minutes, the same amount of LTD as before. With dopamine added to the potassium solution, however, there was LTP (dopamine with the high-frequency depolarization had no effect). So, in this case, pre- plus post-synaptic depolarization by itself produces depression but with added dopamine produces strengthening.

Three rules – several learning patterns

So far, we have three basic learning mechanisms, in terms of mental mechanics: LTP, ADF, and ADF-dependent LTP (ADF➔LTP). LTP – which you can think of as pure association – can account for classical conditioning (with stimulus substitution) and for sensory pre-conditioning. As we will discuss shortly, it can also account for perceptual learning. ADF – which you can think of as pure reinforcement – can account for classical conditioning (with or without stimulus substitution). ADF➔LTP can account for instrumental conditioning. The different aspects of these mental mechanics are illustrated in Figure 6.7.

You may already have noticed that we have two distinct ways of accounting for classical conditioning with stimulus substitution. So, seeing that kind of conditioning at the behavioural level does not immediately tell you which mechanism is involved in any particular case. There are also issues with saying, simply, that classical conditioning results when "cells that fire together wire together". Often the stimuli are separated by a few seconds. Indeed, as I

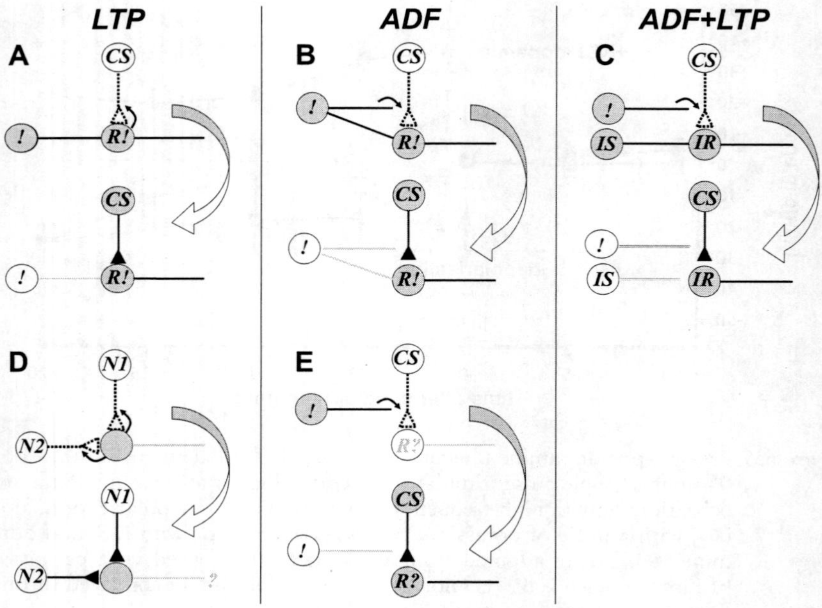

Figure 6.7 Three rules – several learning patterns. (A, B) LTP (A) or ADF (B) producing classical conditioning with stimulus substitution. (C) ADF-dependent LTP producing instrumental conditioning. (D) LTP producing sensory pre-conditioning or perceptual learning. ? = response output after learning of a conditioned response to the neutral stimuli N1 or N2. Combination of weak N1 + weak N2 can fire the post-synaptic cell (grey fill). (E) ADF producing classical conditioning with production of a new response (i.e., without stimulus substitution) not controlled by the experimenter (?). O = cell body (grey fill when active); —— = axon (black when active); △ = weak synapse (black fill when strong); CS = conditioned stimulus; IS = innate stimulus; R = response; ! = unconditioned stimulus that produces unconditioned response (R!).

noted already, the rule derives from perceptual learning (i.e., learning to see) where things are more simultaneous, as we will discuss in the next section. Further there are cases of what looks like instrumental conditioning that, if you look carefully, actually use a classical mental mechanism (particularly with 'autoshaping').

So, we need to apply our mental mechanics carefully, and test in each case which mechanism produces which behavioural change. That will be the business of the next few sections.

The misbehaviour of organisms

Instrumental conditioning procedures can produce the right response for the wrong reason or a wrong response. In both cases, the intended instrumental learning is impacted by incidental classical learning.

My first example is of pigeons 'autoshaping' to press a key. Autoshaping (automatic shaping of responses) is convenient. 'You' train responses without sitting there all day pressing a button to deliver food when the desired behaviour occurs. Let's look at this sequence of multiple-choice answer experiments. My own answer is in the endnote at the end of each question.

Experiment 1

A pigeon is placed in an operant conditioning chamber. After it has settled down, you start lighting one of the keys on the front panel for a few seconds. As the light switches off, you deliver a food pellet. After a number of trials, the pigeon will start pecking the key when it is lit (this is the autoshaping). You observe the sequence: light on – pigeon pecks – pigeon gets food. Is the observed change in behaviour the result of[14]

a classical (respondent) conditioning
b instrumental (operant) conditioning
c both
d not sure

Experiment 2

A pigeon is autoshaped as in Experiment 1. But now you add an omission contingency – that is *if it pecks* the key the food is *cancelled*. Here a positive classical conditioning contingency (light-food, which normally causes a response) is pitted against a negative operant contingency (which normally suppresses responding). Pigeons, usually, cycle between responding (which cancels food) and not responding (which delivers food). *Given this*, do you think pigeon *operant key pecking* normally depends on underlying processes of[15]

a classical (respondent) conditioning
b instrumental (operant) conditioning
c both
d not sure

Experiment 3

A rat is tested in essentially the same manner as the pigeon with autoshaping with omission – the rat generally fails to respond, and just collects the food whenever it is delivered. Given this do you think rat *operant lever pressing* normally depends on underlying processes of[16]

a classical (respondent) conditioning
b instrumental (operant) conditioning
c both
d not sure

A question you can immediately ask is whether this shows that rats are smarter than pigeons. However, you should note that in the pigeon case, they are being asked to *peck* the key. Pecking is what they do 'instinctively' to eat food, and so this could encourage classical stimulus substitution. Indeed, if you look at pigeon key pecks carefully, they have a different form depending on whether a thirsty pigeon is rewarded with water, or a hungry pigeon is rewarded with food. Essentially, you can see they are drinking or eating the key, respectively (Figure 6.8).

The rat example does not use an instinctive response. Rats pick up and chew food; they don't press it like a lever. To see some rats getting confused between producing an instrumental response (a slam dunk) and a classical response (fondling and chewing the ball) go to YouTube and watch "Rat basketball at Wofford College" – and try not to laugh too much if there are others close.

A larger number of different problems with such interfering instinctive responses were described by Keller and Marian Breland.[R3] They were

Figure 6.8 Pigeon operant pecking. Key pecking is stereotyped with successive response forms very similar for any particular reward. On the left, a pigeon is pecking for WATER when thirsty. On the right, the same pigeon is pecking for FOOD when hungry. Images from Jenkins and Moore[R11, their Fig. 2] used with permission of John Wiley & Sons; permission conveyed through Copyright Clearance Center, Inc.

'behavioural engineers' interested in producing displays using animals as convention exhibits, tourist attractions, and television commercials. In the process, they conditioned nearly 40 species (and over 6,000 individuals). They list a large number of varying examples, all with the same kind of intrusion of untrained instinctive behaviour – but with each example depending on the species. One comparison should be enough here. Both involve training to put a coin-like object into a piggy bank. A raccoon (which has hands) could learn to pick up the coin and, initially, put it into the box but soon, and especially when required to put in two coins to get the reward, he would spend a long time rubbing the coins together and just dipping them into the coin slot; and this got worse with time, reducing reward deliveries. A pig (with large wooden coins and a matching piggy bank) learned to put coins in the box rapidly and neatly; but, over a period of weeks, would run eagerly to get the coin but go slower and slower to put it in box – dropping it and rooting it repeatedly. Note that for the raccoon, 'washing behaviour' of the type that would remove the hard shell of a crayfish is an important part of their normal food getting – while, for the pig, 'rooting' is major. At this point you might want to revisit Wofford College, pausing and rewinding at appropriate points to take in the rat's expressions.

Learning to see

A last twist on the learning front is the idea that we 'learn to see'. "Cells that fire together wire together" was coined in the context of visual development.[R18] As we will see in the next chapter, this kind of simple association needs a bit more dynamic mental mechanics if it is to properly explain classical conditioning. But, all by itself, it can explain one of the mysteries of brain development.

> An adult human brain has more than 100 billion neurons. They are specifically and intricately connected with one another in ways that make possible memory, vision, learning, thought, consciousness and other properties of the mind. One of the most remarkable features of the adult nervous system is the precision of this wiring. No aspect of the complicated structure, it would appear, has been left to chance. The achievement of such complexity is even more astounding when one considers that during the first few weeks after fertilization many of the sense organs are not even connected to the embryonic processing centers of the brain. During fetal development, neurons must be generated in the right quantity and location. The axons that propagate from them must select the correct pathway to their target and finally make the right connection.
>
> How do such precise neural links form?
>
> (Shatz[R18, p. 61])

We already have part of the answer from the mechanisms described in Figure 4.3. There we see massive fibre tracts threading their way to their targets using simple (usually two-dimensional) chemical gradients that can be coded very simply. But, what about the detailed wiring once they get to the target? Well, at least part of the answer is, firing together – a process that depends on the environment to provide massive amounts of information that guide development. This has been extensively studied in the developing visual system.

Even before the light-detecting rods and cones have developed, ganglion cells in the retina project to visual thalamus (lateral geniculate) and can generate spontaneous bursts of action potentials. These bursts last seconds and are followed by much longer pauses – with the activity of neighbouring cells more closely linked than cells that are further away (or in the other eye). The terminals releasing transmitter together in the geniculate fire the cell and so tend to wire together (while ones out of synch lose their link) and so the geniculate segregates into left- and right-eye layers (Figure 6.9). The activity-dependence of this is shown by experiments where firing is blocked with tetrodotoxin (that's the poison you need to be careful of if you eat puffer fish) and the wiring stays unlayered.[R18]

The same "cells that fire together wire together" and "cells out of synch lose their link" processes are important for the wiring of the cortex but, in

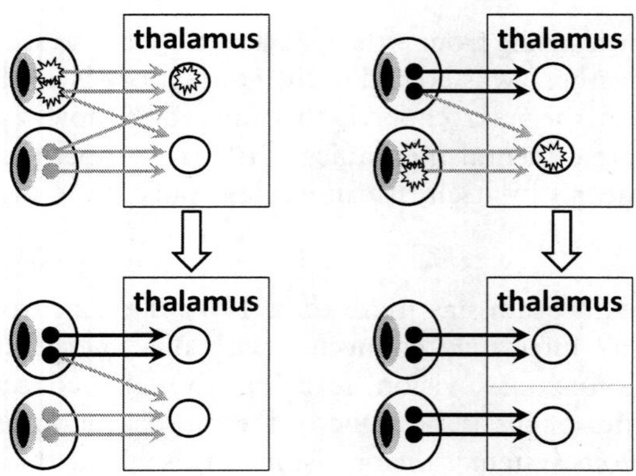

Figure 6.9 Pre-natal thalamic wiring. Although there are no light-sensing cells in the retina, ganglion cells (●), with weak connections (➔) to the lateral geniculate nucleus of the thalamus, fire (❋) in occasional, synchronized, bursts. The coincident input from this eye to the thalamus causes the thalamic cells (○) to fire and pre- and post-synaptic firing together wires them together (●➔). Inputs from ganglion cells in the other eye, which are not firing, are out of synch and lose their link. Output from the active eye to thalamic cells that are not firing is not changed. Separate burst firing in the other eye wires and unlinks other connections in the same way. The end result during development is left and right-eye layers in the thalamus. For more details on this, see Shatz.[R18]

this case, are driven by post-natal experience. This re-wiring is affected by the content of experience, but the simplest example is of what are called ocular dominance columns in the visual cortex. These are adjacent columns of cortical cells where one will receive input largely from one eye and another column from the other eye.

Before visual experience, the inputs from the two eyes have largely overlapping territories (Figure 6.10A), with their location driven by things like chemical gradients (Figure 4.3). Normal visual experience (Figure 6.10B), driven by the closer correlation of firing of cells driven by one eye than cells driven by different eyes, causes a concentration of eye-specific terminals to form within a specific zone and for terminals linked to the other eye to shrink within that zone. While this is going on, more terminals are forming so inputs are not only more efficient (through LTP) but more extensive within their specialized zone. As with the pre-natal wiring of the thalamus, this post-natal segregation can be blocked with tetrodotoxin (not shown) although the expansion of terminals is not. If one eye is blocked with a patch in this critical developmental period, the closed eye loses territory and the

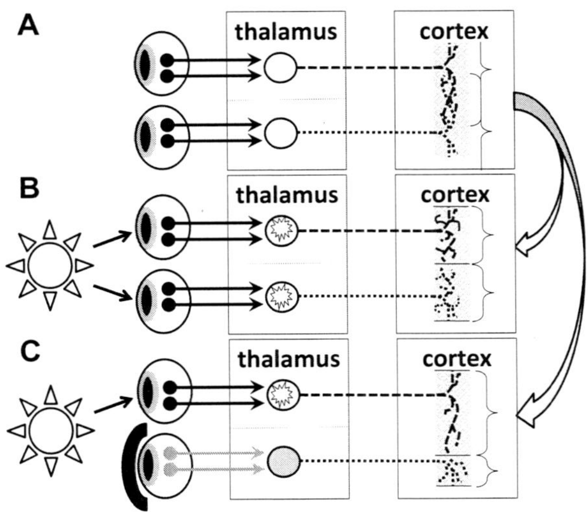

Figure 6.10 Post-natal cortical wiring. (A) Thalamic cells (O) initially send widespread, overlapping connections to cortical cells. (The overlap between left- and right-eye cells is shown, with the territories of each extending into the cell layer above and below the area shown. (B) When both eyes receive images, the thalamic cells fire (❋). Their input to the cortex is more correlated within an eye than between them and so the territories of the inputs from the two eyes shrink and separate to form discrete adjacent columns of cortical cells ('ocular dominance columns'). (C) If, instead, input to one eye is blocked, for example, with a patch, so that it never receives images, its territory is greatly reduced, while the input of the other eye is strengthened, taking over the territory that the blocked eye would have had. For more details on this, see Shatz.[R18]

open eye keeps the territory it had at the beginning of development and does not shrink (Figure 6.10C).

From learning to memory

There is a sense in which, once you have learned to see, your brain has a long-term memory of how to do it. Of course, this is not exactly the same as what we usually call memory – nor, for that matter, is remembering how to ride a bike. To fully understand classical and instrumental conditioning under real-world conditions (including the simple demonstrations in the laboratory), you need to understand what is often called working memory (though I prefer to call it active memory following Fuster[R9]).

So, learning and memory are tightly tied together. But they feel like quite different things. So, we will look at memory largely separately in the next chapter – cross-referring where necessary.

Notes

1 For an example of how to do this in the laboratory using a 'Visible Burrow System', see Blanchard and Blanchard.[R1]
2 See 'Sandra's story' in Chapter 10 for a direct quote.
3 See Chapter 5, especially *Emotion and feelings*.
4 The story of the discovery of LTP by Terje Lømo, and related trials and tribulations in later work on it, is told by Tim Bliss.[R2]
5 N-methyl-D-aspartate receptors, for which the transmitter is glutamate; these are different from α-amino-3-hydroxy-5-methyl-4-isoxazolepropionic acid (AMPA) receptors, for which glutamate is also the transmitter. The full names for NMDA and AMPA are seldom written out.
6 Note * and italics in section heading – this one is highly technical.
7 See Chapter 3.
8 Hebb lived long enough for him to be told about the associative properties of LTP by one of his 'great-grand-students' but saw his cell assembly and phase sequence ideas as more important.[R14]
9 For details of older variations of this idea, see Brown et al.[R5] They emphasize that Hebb added the concepts of the cell assembly and phase sequences in such assemblies to connect the synaptic level to mental-level concepts – or what Hebb called the Conceptual Nervous System.
10 Although Pavlov's 1927 book is often cited for the origin of the phrase 'stimulus substitution', he does not appear to have used the term himself but says, instead, "It is obvious also that the underlying principle of this activity is signalization. The sound of the metronome is the signal for food, and the animal reacts to the signal in the same way as if it were food; no distinction can be observed between the effects … of the beating metronome and … real food".[R15, p. 22]
11 Especially in the army, training very often involves omission of an expected punishment. In terms of learning processes this acts functionally like a reward. Conversely, omission of an expected reward produces frustration that is functionally equivalent to presentation of a punishment. Omission contingencies are harder to get your head around than simple presentation, and so I have stuck with reward here to keep things simple. For more on this and the distinction between valuation (gain/loss) and motivation (approach/avoid) see Corr and McNaughton.[R7]

12 I have a conflict of interest here as the work was done at Otago, and they are all friends of mine.
13 Complicated compounds made and supplied by drug companies often have simple letter plus number labels like this. SCH stands for Schering (the pharmaceutical company that produced it) and the number is their index number for it. It is a halobenzazepine: (R)-(+)–7-chloro-8-hydroxy-3-methyl-1-phenyl-2,3, 4,5- tetrahydro-1H-3-benzazepine. No wonder it's always referred to by the number. To give a slightly different example, Ro 15–1788 was produced by Roche and is chemically 5,6-dihydro-4H-imidazo[1,5-a][1,4]benzodiazepine. It is used clinically and so usually referred to by the name flumazenil. But it can also be called Anexate, Flumazepil, Romazicon and a few other names. So, with most such drugs, in the lab, the letter-number code usually seems safer.
14 Your most likely initial guess is probably (b); my own choice is (d) for reasons that should become clear as we progress through the experiments.
15 Here I would expect you to choose (a). Light food is causing pecking – we have stimulus substitution since pigeons usually peck food. Critically, the omission contingency breaks the classical light-food pairing and so results in extinction of the response.
16 Here I would expect you to choose (b).

References

R1. Blanchard, RJ & Blanchard, DC, Antipredator defensive behaviors in a visible burrow system. *Journal of Comparative Psychology*, 1989. **103**: 70–82.

R2. Bliss, T, Pursuing synaptic plasticity from cortex to LTP in the hippocampus. *Hippocampus*, 2025. **35**: e23665.

R3. Breland, K & Breland, M, The misbehaviour of organisms. *American Psychologist*, 1961. **16**: 681–4.

R4. Brogden, WJ, Sensory pre-conditioning. *Journal of Experimental Psychology*, 1939. **25**: 323–32.

R5. Brown, RE, Bligh, TWB, & Garden, JF, The Hebb synapse before Hebb: Theories of synaptic function in learning and memory before, with a discussion of the long-lost synaptic theory of William McDougall. *Frontiers in Behavioral Neuroscience*, 2021. **15**: 732195.

R6. Carew, TJ, Hawkins, RD, Abrams, TW, & Kandel, ER, A test of Hebb's postulate at identified synapses which mediate classical conditioning in Aplysia. *Journal of Neuroscience*, 1984. **4**: 1217–24.

R7. Corr, PJ & McNaughton, N, Neuroscience and approach/avoidance personality traits: A two stage (valuation-motivation) approach. *Neuroscience & Biobehavioral Reviews*, 2012. **36**: 2339–54.

R8. Edwards, S & Uhlenhuth, EH, Panic disorder and agoraphobia: A sufferer's perspective. *Journal of Affective Disorders*, 1998. **50**: 65–74.

R9. Fuster, JM, *Memory in the cerebral cortex*. 1995, Cambridge: MIT Press.

R10. Hebb, DO, *The organization of behavior: A neuropsychological theory*. 1949, New York: Wiley-Interscience.

R11. Jenkins, HM & Moore, BR, The form of the auto-shaped response with food or water reinforcers. *Journal of the Experimental Analysis of Behavior*, 1973. **20**: 163–81.

R12. Kandel, ER & Hawkins, RD, The biological basis of learning and individuality. *Scientific American*, 1992. **267**: 79–86.

R13. LeDoux, JE, Emotion, memory and the brain. *Scientific American*, 1994. **270, June**: 50–9.

R14. McNaughton, BL, Long-term potentiation, cooperativity and Hebb's cell assemblies: A personal history. *Philosophical Transactions of the Royal Society B: Biological Sciences*, 2003. **358**: 629–34.

R15. Pavlov, IP, *Conditioned reflexes (trans. G. V. Anrep)*. 1927, London: Constable & Co. Ltd, Dover edition, 1960, by special arrangement with Oxford University Press.

R16. Prewitt, EP, Number of preconditioning trials in sensory preconditioning using CER training 1. *Journal of Comparative and Physiological Psychology*, 1967. **64**: 360–2.

R17. Reynolds, JNJ, Hyland, BI, & Wickens, JR, A cellular mechanism of reward-related learning. *Nature*, 2001. **413**: 67–70.

R18. Shatz, CJ, The developing brain. *Scientific American*, 1992. **267**: 60–7.

R19. Wickens, J, Begg, AJ, & Arbuthnott, GW, Dopamine reverses the depression of rat corticostriatal synapses which normally follows high-frequency stimulation of cortex in vitro. *Neuroscience*, 1996. **70**: 1–5.

7 How do we remember?

What is memory?

I just argued that associative change in the visual system during development is 'learning to see'. From that perspective, your visual system remembers what it learned: i.e., how to see. But the system 'remembers'; it does not have a specialized 'memory'.

The other types of associative learning also produce long-term memories: that food or shock is coming, or that you should press a button to get food. That said, the 'perceptual memory' that you have of, for example, your grandmother's face is obviously different from the reportable memory that a light signals food or the unreportable memory of *how* to ride a bicycle (as opposed to *that* you can ride a bike). These types of long-term memory are also clearly different from what is usually called working memory. I prefer to call it active memory following Fuster,[R5] when I can.

Donald Hebb[R7] suggested that perceptual memory involves 'cell assemblies' that code not only the fundamental aspects of memories but also, at the first instance, the original perception. These can be activated by external input (ultimately producing perception) or internal association (long-term memory). In both cases, they can then be kept cognitively active by recycling activity in networks of neurons. Recycling keeps the engram 'in mind'. Like long-term memory, there are different sorts of active/working memory – which often work together. In addition, as we shall see, long-term classical conditioning of simple associations (LTP) may need to involve active memory under normal laboratory testing conditions (and normal living).

Cell assemblies can be simple, e.g., coding the position of a dot on a screen. But they can be complex – coding faces and words and episodes. The eye does not produce a digital photo image and send it up to the cortex. This is too much information and too undigested to guide choices. So, information is compressed and converted to a set of parameters based on a template. These can identify an individual, and we will discuss such perception as a whole in Chapter 9. For now, you just need to remember that both simple and complex percepts can be held in cell assemblies that code the engrams used in memory.

DOI: 10.4324/9781003465027-7

Finally, note that 'declarative'/event memory results from a single exposure, not repeated training. It involves much more information than simple tones and lights; and it clearly depends on some kind of association, not on reinforcement. You did not get rewarded at the time for remembering where you had lunch yesterday. We will discuss the issue of the strengthening (consolidation) of such memories in the broader context of sleep in Chapter 8.

These later issues will all take some unpacking. What is active (working) memory? It has distinct types. We need to understand how classical and instrumental conditioning generate learning under real-world conditions.

What is active (working) memory?

To answer this question, let's start off with Experiment 7.1.

Experiment 7.1 Active memory versus reference memory

To check this out, you will need a volunteer friend or relative as a participant.

Step 1: ask them to tell you their phone number. If they are an oldy like me ask them for their home (landline) number as they likely cannot remember their cell phone number (so I have mine written inside my cell wallet). They should manage this quickly and correctly.

Step 2: Ask them to listen carefully to the following series of numbers that you will read out:

479700 474369 936700 558416 496402 582592

Step 3: ask them to recall the numbers, in order. You will usually find they do this slowly, and can manage the first one, sometimes the last, but seldom more than a couple (unless they have a super power number memory) and usually with some mistakes. As it happens, all of these numbers are truncated parts of phone numbers that many people in a big city in New Zealand can remember straight off – so it's not that they are difficult to learn.

Step 4: Ask them to listen to the following number, then wait until you ask them to tell you what it is.

640258

Step 5: Count to 30 slowly (this should be quite a bit longer than it took you to read the entire first set of numbers) or time a longer interval if your participant is happy to try that. Then ask them to tell you the number. They will almost certainly respond quickly and get it completely right. Ask them how they did this, and they will say they repeated it over and over in their head. The new number is actually the middle segment from the last two numbers of the previous list (49 640|258 2592) so it should have been just as easy or difficult to remember as a number.

You have just demonstrated that we can hold some numbers in a long-term (reference memory) store. But as you will know from when you got a new phone number, this takes a while to learn. It also shows that we have a short-term limited capacity store. You can hold a number in your mind provided you keep it active – but you can't do that with a lot of numbers.

What is the mental mechanics of active (working) memory?

Donald Hebb (who we met in relation to LTP in Chapter 6) provided a simple theory of how active memory works. His idea was that information can be held, actively, in reverberating circuits (Figure 7.1). The simplest case involves what he called cell assemblies. We will look at these in more detail later, but for our current purposes, the point is that you need sufficient cells linked together in repeating circuits to allow activity to continue for a time. If you simply had two cells, activity would die very fast as the first cell would still be recovering from firing (it would be in what is called its refractory period) when the second cell tried to send the activity back. With a typical cortical cell assembly, containing multiple cells, Hebb thought that activity could loop *within the assembly* for up to about ten seconds ("the apparent duration of conceptual processes" in humans[R7, p. 143]). The activity has to be set to decay eventually or the result would be runaway activity and brain seizures.

In Figure 7.1, this is shown as a brief visual stimulus, relayed by the thalamus, activating the first cell of a cortical cell assembly that codes for that stimulus (Step 1). This first cell then activates the second, which activates the third, and so on, repeatedly (Step 2).

How does active memory support classical conditioning?

Output from any of these cells can then be used for any process that needs information about the stimulus when the stimulus has disappeared. As an example, I have shown the operation of active memory during classical conditioning based on LTP (and so involving stimulus substitution as explained in Chapter 6).

Something I did not emphasize in Chapter 6 is that LTP occurs only with essentially *coincident* pre- and post-synaptic activity, but classical conditioning, as in the Ledoux example, can occur with a separation of the CS from the US of several seconds. How can the CS and US cells wire together if they don't fire together? The answer is that more cells, in an assembly, are involved.

As explained in Figure 7.1, reverberation in a cell assembly can bridge this gap. The CS may have stopped several seconds ago but activity coding the CS will still be circulating. If one of the bursts of weak output from the cell assembly to a cell coding an UR coincides with that cell being fired by the US (Figure 7.1, Step 3), it can be strengthened via LTP (as described in Chapter 6).

How does active memory support delayed responding?

In addition to reverberation within a cell assembly, activity can circulate in longer loops between distinct areas of the brain. A stimulus can activate, say,

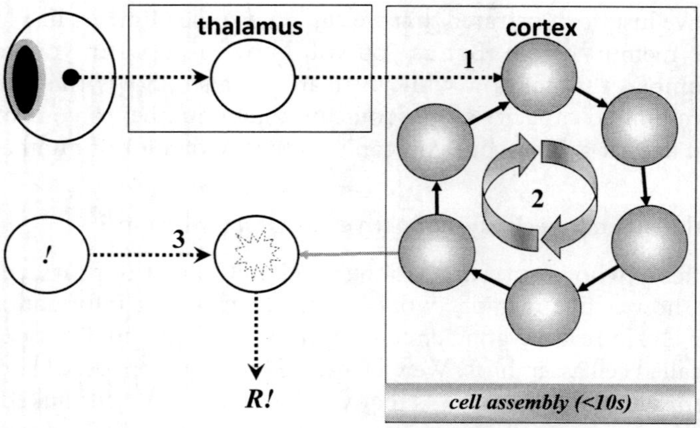

Figure 7.1 Hebb's view of reverberating circuits and active memory. A **brief** stimulus enters the eye and is processed in the thalamus which then (dashed arrow Step 1) sends a signal on to one of the cells in a cortical cell assembly that codes the stimulus. Activity then circulates via the connections between the cells of the assembly (➔), and so the signal is maintained, in the cell assembly (Step 2), but will fade over less than ten seconds. This can explain classical conditioning with delays of up to ten seconds. If an unconditioned stimulus (**!**) is presented, it will **briefly** activate (Step 3) the cells controlling the unconditioned response (**R!**). Provided that this activation overlaps the period of reverberation in the cell assembly (❋) this will strengthen the weak connection (➔) from the cell assembly – as described in Chapter 6.

visual cortex. This activity then passes to and is returned by prefrontal cortex, so keeping the cell assembly active (somewhat like the verbal rehearsal used in Step 4 of Experiment 7.1 but using a different mechanism as we will see in the next section). This can continue for a long time (provided it's not interrupted) and make information available to guide a future action.

Looping was recorded by Patricia Goldman-Rakic at the level of single cells in monkeys carrying out a visual target task.[R6] A target dot was shown briefly at the side of a screen. Meanwhile, the monkey had to stay fixated on the centre of the screen. After a delay a 'go signal' was presented and the monkey had to move its eyes from the centre to the position that the dot had occupied. During this delay frontal and visual cells kept each other active – retaining the dot's position information and allowing accurate responding. The 'go' signal released the response and then cancelled this looping. This type of prefrontal-perceptual cortex loop activity, holding information 'in mind' temporarily, is an integral part of the control of ongoing motor acts. That is one reason why most people prefer the term 'working memory'.

How many components of working (active) memory are there?

Alan Baddeley has spent many years evolving a Multicomponent Theory of working memory, see Baddeley[R2] for overview. All of his components employ looping – but of different sorts of information in different ways. These components and the relation between them are shown in Figure 7.2.

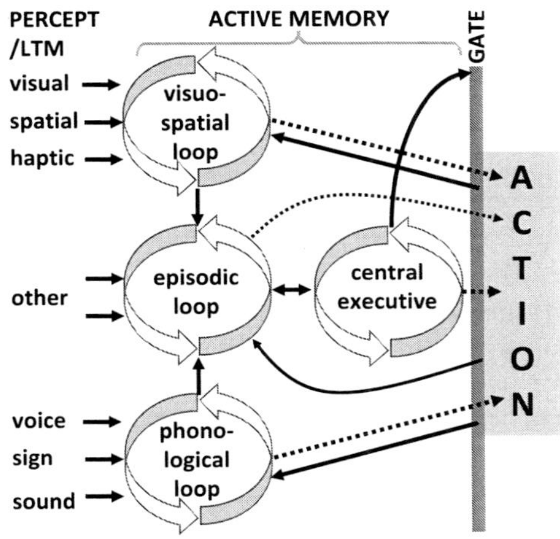

Figure 7.2 Types of working/active memory. These all keep information in cell assemblies active. The information can have started as novel input (percept) or be held in the cell assembly by long-term changes in synaptic strength that result in an engram in long-term memory (LTP). **The visuo-spatial loop** (or 'sketch-pad') is the one used in Goldman-Rakic's monkey visual target experiment. It processes visual, spatial, and haptic (touch and proprioception) information. **The phonological loop** is the one your participant used for the final steps in Experiment 7.1. With spoken words (including numbers) it's the easiest one to experience the looping as you feel as if you are repeating the words inside your head (or often out loud). **The episodic loop** (or buffer) deals, essentially, with all other information – particularly high-information episodic memory. **The central executive** controls attention and, via inhibition, the readout from the other active memory systems to the generation of action. This can be thought of as controlling a gate for that information. Based on Goldman-Rakic[R6] and Baddeley[R2] and adapted from Figure 11.6 of McNaughton and Gray[R9] with permission via author's contract.

The important point is that, for each of the loops – visuo-spatial, episodic, and phonological – a cell assembly in a perceptual area is activated. This activation can be a percept if triggered by external input. But the same assembly can be activated by associations as recall of a previous percept from long-term memory. The looping keeps the assembly active – and so the engram that the cell assembly codes is kept actively in mind. From an evolutionary point of view, the purpose of such keeping in mind (whether of immediate or old information) is to guide current action. It is the business of the central executive to control action sequences and the contribution of the other loops. In particular, it prevents action that must be delayed (as when the monkey is waiting for the go signal in the visual target case above).

Is there a long-term memory store?

Donald Hebb's idea of the cell assembly[R7] saw its re-activation as the basis for retrieval from long-term memory. That is, the long-term storage of the engram involved both strengthened connections between the components of the cell assembly (storing the engram) and strengthened associative connections of the assembly to other cell assemblies. Given such strengthening, activating any of the other linked assemblies would re-activate the engram (in approximately the same configuration as the original percept) and so retrieve the memory.

On this view, perception, retrieved memory, and imagination of a particular percept would all involve activation of the same cell assembly, and so the same part of the brain. Importantly, this means that if you tried to delete the memory you would also make the person blind to the equivalent percept – and so you would have no way of testing some other possible location of the memory since its perception had been destroyed. This explains the results obtained many years ago by Karl Lashley.[1]

> Lashley, the most influential physiological psychologist of his era, had spent 35 years (1915–1950) training rats, cats and monkeys to perform complex learning tasks and then cutting, destroying, or removing specific parts of their brains in a vain attempt to erase the memory of their training. For example, in one series of studies, rats received lesions of various sizes to different parts of the cerebral cortex after they had learned a maze task. Ten days later, their retention of the task was assessed. Lashley found that only large cortical lesions disrupted retention and that the particular site of the lesion was of little consequence: Cortical lesions of equal size produced similar effects regardless of their location.
>
> (Pinel[R17, p. 372])

Although care must be taken in interpreting Lashley's experiments, the results go against the idea that the brain has a specific separate memory store. But, especially with our frequent (incorrect) comparisons of the brain with a computer, it is tempting to think there must be a store. When I was a student, this store was often seen as being in the hippocampus. As time has gone by, true long-term storage has become allocated to cortex – but many still see intermediate storage (before memories become fully consolidated and so 'permanent') as in the hippocampus. (We'll come back to this in Chapter 8.)

A major reason to see the hippocampus as a memory store is the case of Henry Mollaison (H.M.). He had parts of his temporal lobes, including the hippocampus removed bilaterally in an attempt to control epilepsy. Immediately afterwards, he appeared to have lost his memory (which is why that operation has not been carried out on anyone else since).

But I will argue that H.M. did not have amnesia (a loss of memory). He had a failure to suppress incorrect memories (hypermnesia).[R10] That is, an excess of retrieved items (via association) interfered with recall of the one correct item. To get a feeling for what this might be like, think of the tip-of-the-tongue phenomenon. You know that you know something, but the harder you try, the less you can bring the answer to mind. Many incorrect items are fighting with the one correct item for access to consciousness. Go away and think about something else for a few minutes and, magically, the correct answer will pop into your conscious mind.

Why should you see the hippocampus as not storing episodic memories? There are several major reasons.

First is that our memory capacity seems larger than many other vertebrates. But while our neocortex is about 150 times the size of a basal insectivore's (relative to subcortex), the hippocampus is only about four times the size.[R19] Given that a memory associates percepts, you would expect the store to expand faster than the perceptual equipment not the other way round. Worse, the part of the hippocampus that has expanded *least* is the part that most people now view as mediating "cognitive functions – particularly spatial memory – whereas [the part that has expanded *most* is] involved in emotional responses".[R20, p. 655]

Second, the hippocampus is an old form of cortex (archicortex) that provided the seed for the expansion of limbic and frontal neocortex. (It appears at around the lamprey level in Figure 4.5.) The lamprey is not renowned for a super-powerful memory.

Third, H.M. did not lose his past memories (except for some that might not yet have been consolidated). He lost his capacity to form new ones. The title of the original publication reporting results from H.M. emphasizes the loss of *recent* memory.[R18] We will look at H.M. in detail in the next sections. He is particularly interesting as a single case that has been studied in ways that allow reliable conclusions to be drawn.

Fourth, some argue that the hippocampus holds temporary memories so that they can be consolidated. But this idea has major logical and technical problems that we will deal with in the next chapter.

An appreciation of Henry Mollaison

Henry Gustav Molaison, aged 60, at MIT in 1986.

Photograph (adapted with permission) and copyright:

Jenni Ogden, author of

Trouble In Mind: Stories from a Neuropsychologist's Casebook,

Oxford University Press, New York, 2012

H.M. is probably the most famous patient ever reported in the literature. His dramatic loss of anterograde memory coupled with his retention of intelligence, made him an irresistible (and amazingly compliant) subject of research. For example, Jenni Ogden

> decided to assess H.M.'s ability to measure time and told him that he would be left alone for a period, and then asked to estimate the time that had elapsed. [She] left the room at 2.05pm and returned at 2.17 pm. When H.M. was asked how many minutes had passed, he replied, without hesitation, '12 minutes; got you there!'. There was a large clock on the wall and H.M. had noted the time when [Jenni] left, and continued to rehearse it while looking at the clock. When asked how many minutes had passed, he simply subtracted the time he was rehearsing from the time shown on the clock. This anecdote illustrates more about H.M.'s sense of humour, his willingness to cooperate with anything experimenters may come up with, and his intact intellect, than about his ability to estimate time without memories!
>
> (Milner[R16, p. 199])

It is obviously non-standard to conduct experiments with N = 1. However, the division between what H.M. could do and what he could not do was stark. Also, it is statistically reasonable to compare N = 1 with N = 10 or so. First, you must match the 10 on important memory-related factors like age. The N = 10 then provides you with an estimate of the standard error for the 'N = 1' group. With serious brain damage, the standard error of the latter is likely to be low – with a floor effect on scores so we are being conservative. So, we should offer great, retrospective, thanks to 'H.M.' for making his unique data, including his brain,[R1] available.

But also, we should thank Henry Mollaison for reminding us that neuroscience is about people. The 'subjects' (or, if you want to be PC, 'participants') are people who volunteer for the common good with a positive spirit. As reported by Suzanne Corkin who spent many years delineating the boundaries of H.M.'s deficits,

> although H.M.'s remote memories do not contain as much detail as those of control participants, they do provide some fodder for his self-concept. Furthermore, beyond those fragmentary memories, he has beliefs, desires and values that are always present. For example, he believes that doing crossword puzzles helps him to remember words and is fun. He is altruistic: when I asked him to tell me about Dr Scoville (with whom H.M. had several appointments before his operation) he said, 'He did medical research on people – all kinds of people. *What he learned about me helped others too, and I'm glad about that.*' His social behaviour is appropriate and courteous. Years ago, when he and I were walking from the MIT Clinical Research Center to the Psychology Department, he would cup my elbow with his hand to guide me down the sidewalk. He

> has high moral standards with respect to right and wrong in his personal conduct, which is consistent with the preservation of his orbitofrontal cortex. He has a conscience. For example, when explaining why he could not fulfil his dream of being a neurosurgeon, he cites the fact that he wears glasses, and that blood might spurt up onto his glasses, creating an obstacle to his vision and causing him to miss his target in the patient's brain, thereby causing the patient harm. In addition, he has good insight into his memory disorder. When I asked him, 'What do you do to try to remember?' he replied, 'Well, that I don't know 'cause I don't remember (laugh) what I tried.' He has a sense of humour, and often makes jokes.
>
> (Corkin[R4, p. 158; emphasis added])

How forgiving is it to say, "what he learned about me helped others too, and I'm glad about that" when the surgeon destroyed his capacity for memory (albeit with the best intentions and an improved clinical outcome)? I am about to move to what may seem a very clinical dissection of 'H.M.'. Given the above, I want to emphasize that I have appreciated his contribution as a person since I started meeting his data more than half a century ago. I also feel honoured to know someone who met and worked with him. Mental mechanics is about people, neurotypical or otherwise.

Does the hippocampus store long-term memory?

Well clearly, no. Even the original publication refers to loss of *recent* memory.[R18] We know H.M. could remember where he lived before his operation but not after, once this had changed.

But how do you test remote memory? You can't go back in time, present a stimulus and then check for recall. I think that the Famous Faces Test is a cute answer to this question. To build this, you pick sets of faces that were famous (and so presented as 'stimuli' by news media) in particular decades. Then you compare current recall of such faces in relation to decade. To see roughly how this works, try Experiment 7.2. The key thing is that you don't have to go back in time; the images do this – if their fame is local to a decade and does not go on and on.

Experiment 7.2: Testing remote memory with famous faces

When used to test remote memory, the Famous Faces Test goes out of date fast. Time keeps marching forward so the faces' decades keep marching backwards. For normal use, it has to be constructed and normed to match the culture of those being tested (USA versus UK need different sets); and, unlike the illustrative ones below, the images would normally not be organized by decade. Your age is also an issue. The version on which H.M. was tested has its latest examples from the 1970s. I am old

enough to know most of the 1960s and 1970s ones. But it's quite likely that you were not even born then, far less interested in famous people.

To try and get around these cultural problems, I built the example below using American Vice-Presidents (who tend to have a fairly broad news coverage around the world). I excluded those who later became president (and so would have had more coverage and for longer than a decade). Even one of the included cases is not ideal as they were famous later for other things – I flag this one in the end note that decodes the faces' identities.

To test yourself:

1. look at the faces, and write down who you think they are – organized left to right and top to bottom like the faces.
2. Then look up the correct answers here in this endnote[2]. Tick the ones you got correct.
3. Now add up how many you got correct for each major time period. If your results are like the controls who were tested against H.M., you will be most frequently correct with the recent faces (many people will get 100%) and least with the oldest ones (I tested a few people my age and they got 0%). You could have fun testing some of your friends, too.
4. As this is a non-overlapping series, you can also check the time effect by looking at your results in terms of left-to-right then top-to-bottom order. Top left is the oldest, bottom right is the most recent.
5. But remember this is just an explanatory example and has not been validated or normed.

When tested on one version of the Famous Faces Test.[R15] HM shows (Figure 7.3, black symbols): intact remote memory (up to about ten years before his operation); retrograde (backwards in time) amnesia that is worse closer to his operation; and anterograde (forward from the operation) amnesia that is worst of all. Particularly interesting are the results from a second presentation of faces that had been missed on the first presentation.

> If on this second presentation the subjects could not identify the face correctly, they were then given either a circumstantial or a letter prompt. An example of a circumstantial prompt (for Alfred Landon) is as follows: 'He was the Republican presidential nominee in 1936; he ran against Roosevelt and lost. He was also governor of Kansas.' The letter prompts were graded systematically. The subject

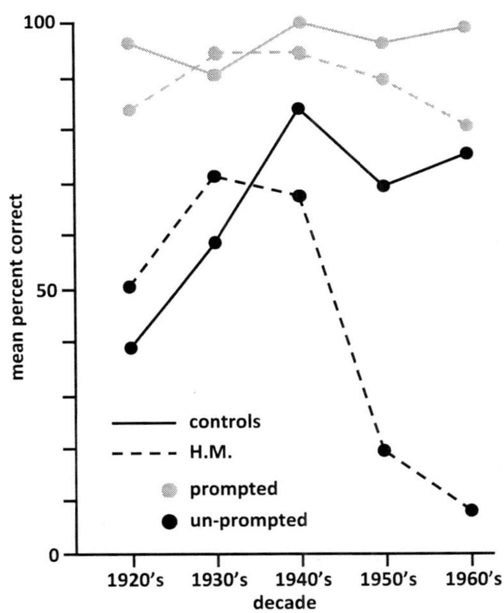

Figure 7.3 Remote memory in H.M. The graph shows the results when testing HM (- - - -) on the Famous Faces Test (see Experiment 7.1). He is compared to a group of age-matched controls (—). Everyone does poorly on photos from the 1920s (HM was born in 1926). But until the 1940s H.M. scores, if anything, better than the controls. His scores then drop down markedly hitting their lowest point in the decade of his operation (in 1953). When they were tested with progressively stronger prompts (●) everyone showed improved scores. But HM shows a much greater improvement and is essentially like prompted controls for the decades before his operation and like unprompted controls for the decade after. Data retrieved from figures in Marslen-Wilson and Teuber.[R8]

> was first given the initials, then the first syllable of the first name, then all of the first name, the first syllable of the last name, and as far beyond that as possible letter by letter, without giving the full word. Thus, for Alfred Landon the letter prompt sequence was as follows: A.L.; Alf. L.; Alfred L.; Alfred Lan...; Alfred Land... If the prompts had not elicited the full name by this time, the subject was given the rest of the last name and asked if he had ever heard of the person in question. We alternated between circumstantial and letter prompts.
>
> (Marslen-Wilson and Teuber[R8, p.356])

As shown in Figure 7.3 (grey symbols), everyone's scores increased with prompting. So, the controls clearly had memories that they were not retrieving in the first test. H.M.'s retrograde amnesia was eliminated and his anterograde amnesia still resulted in scores fractionally better than the *unprompted* controls. The simplest explanation of this is that prompting (like the waiting period for the tip-of-the-tongue phenomenon) reduces interference from competing alternative engrams (most of which will not match the prompts).

Intact procedural and perceptual learning in 'amnesia'

The results with the Famous Faces Test suggest that (whether it is a true memory problem or something else), H.M. had major problems with storing new information. However, it turns out that this only seems to be extreme with declarative memory (the kind of memory where you can report the content of what you have learned). Here we will look at H.M.'s procedural and perceptual learning.

You can test procedural learning with Mirror-drawing. This has the advantage of being something we do not normally have any experience with. In the original test, a mirror is set up in front of you, and you have to look into it to see an image that you then have to trace – but without being able to see it directly. To get an idea of how it works (but using a computer camera to get the mirror effect) try Experiment 7.3. When you have tried the experiment, ask yourself if you could simply be instructed how to do it and then get it right the first time. (It usually feels fairly frustrating when you start the test and don't get it right.)

H.M. showed normal learning and excellent retention of mirror drawing over a three-day period (Figure 7.4). But he could not recall that he had done the test before.[R11, cited by R12] He showed a similar capacity to learn other manual tracking and coordination tasks; with limited retention of simple mazes and, like other amnesics, a failure to learn sequences greater than his active/working memory span.[R12]

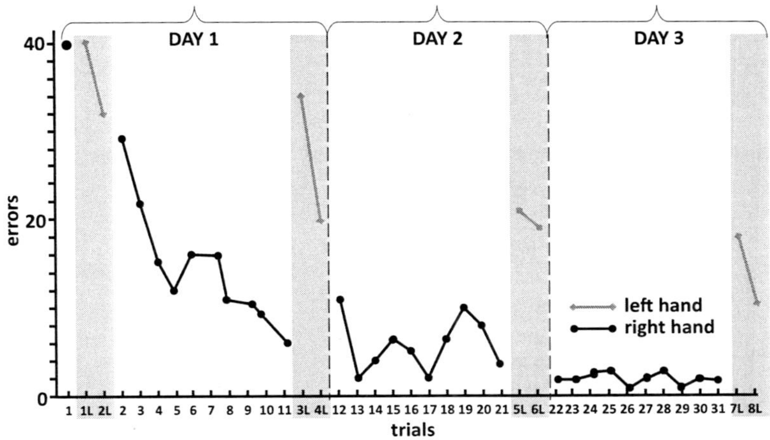

Figure 7.4 Mirror drawing by HM. For basic details of the test, see Experiment 7.3. Occasional tests of the left hand (grey) were interpolated into a series of trials with the right hand (black). The first of these was immediately after trial 1. Note both the steady learning and the strong retention of the learning between days. If you do Experiment 7.2, see if you can get better scores than H.M. Redrawn from Milner.[R12]

Experiment 7.3 Mirror drawing

The original test uses a real mirror held at an appropriate angle in front of you. I didn't have a movable one in the house. I presume you won't either. I also had visions of broken mirror glass! So, here is a workaround using a laptop computer with a plug-in camera. (You can't get the right angle with the built-in camera.)

You need to set up your laptop on a table or desk that you can sit at. You may be able to perch the camera on the screen. But I found it much easier to hold the end of the camera mounting in a bulldog clip, and then tape one clip handle to the underside of a box of tissues. This means you don't get sticky tape in places you don't want and can adjust the height by putting a box, or a pile of books, under the tissue box. The camera mounting also provides more flexibility. This set-up is shown in the photo below.

You now need to set up a cardboard shield (as at right in photo, tilted a bit towards the computer) to block your view. With this, you can draw on the stars but not see them, except on the computer screen. Start with a 12-bottle wine box (or similar). Slit the tape on the underside so you can open the box out into its original flat pack. Fold the

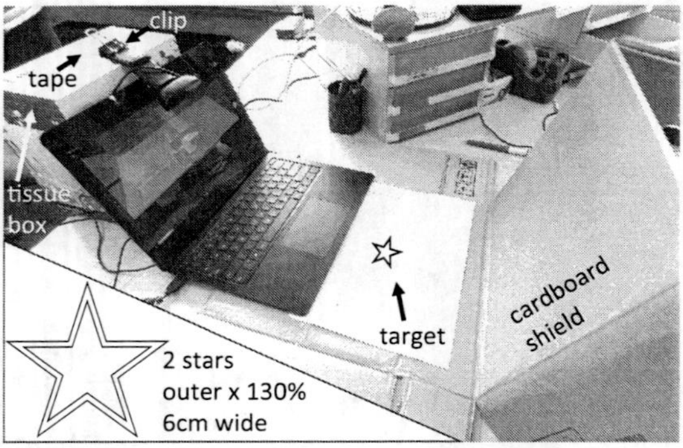

flat pack up at the join between the uppermost long and short side. (This will fold the underside too but not at its join, see bottom right of photo.) Then, fold the lid sections of the smaller side outwards. You will have to cut the lower, attached, ones free with a pair of scissors or they will just stay sticking out and get in the way of your arms. Place the shield just in front of where you will be seated, with the long side flat and the short side providing the shield. Place the laptop on it (as in the photo), leaving enough space for an A4 sheet of paper.

You then need to make a five-point double star (as shown below left of the photo) and print it. You can start with my star and scan (or photo) it, paste it into an A4 page, and enlarge it to 6 cm wide. But you will get a neater result if you draw it in a simple drawing program. For PCs you can draw the stars as an object with Paint or PowerPoint. (PowerPoint gives you more control.) For MACs you can use Pages. My stars are scaled to each other as in the image in Milner[R12] with the outer copied at × 1.3 of the inner. Keep the line thickness fairly fine as it's the white area between the two that is the target in the test. You will need to print 38 copies for each person who wants to do the full test.

Place the sheet of paper with the star just in front of the laptop in a position where you can easily draw on it (see photo). Start your camera program (for MAC, open Photobooth; for PC – open the inbuilt Camera app). It is likely to default to the built-in camera, so switch to the plugged-in one. Select full-screen mode. Adjust the camera position so you can see the star clearly on the computer screen. Make sure you have a pencil (or ball-point pen) to hand. You can now start testing.

If you want to analyse your results, number your star sheets as per the trials axis in Figure 7.4. This will allow you to keep track as you go; and not have to worry about getting their order mixed for analysis. Place Sheet 1 in position so it's in the middle of the computer screen. Sit down with your arms round the sides of the shield so you can hold the paper down with one hand and draw on it with the other. Make sure you can see the star on the screen but that both your hands and the star are blocked by the shield. (This may want to be tilted towards the laptop a bit.)

Starting at the top point of the star, trace in between the two lines using the hand you normally write with until you have completed the star. If you cross either of the lines, that will be counted as an error. Replace Sheet 1 with Sheet 1L, switch the pencil/pen to the other hand. Trace the start. Repeat with Sheet 2L. Shift back to your dominant hand and repeat with this for Sheets 2–11. Switch hands and do 3L and 4L. If you want to test longer-term memory, test yourself again over the next two days using the trials sequence shown in Figure 7.4.

After you have completed a day's testing, make sure the sheets are in order and score each one. Count every line crossed from inside to outside, either double line as an error and write the total score on the sheet. The score on trial 1 is simply a test of your current mirror control and does not involve any learning. H.M. made 40 errors with his right hand and then the same on trial 1L with his left hand. The key result for memory is whether this value goes down with later trials and stays down over days (see Figure 7.4). See if your learning and memory are better than those of the 'amnesic' H.M.

Perceptual memory is another area where H.M. showed largely normal learning and retention. Again, it appeared important that the core of what was being learned was hard to verbalize. The incomplete pictures test presents sets of partial pictures (e.g., the degraded parts of the inset plane in Figure 7.5). For initial learning, sets of progressively less incomplete images are shown until the identity of the complete image is recognized. After 1 hour the procedure is repeated.

As you can see in the figure, the pattern of H.M.'s recognition of the initial partial pictures was essentially the same as that of controls who had been matched with him for initial performance. After 1 hour, control recognition on the first (most partial set) showed average savings of ~75% (range 52%–93%). H.M. showed clear savings (48%) just below the bottom of the control range. He reached 0 errors on Set 3, as did controls.

There are other examples of H.M.'s residual (but poor) learning and memory retention over days in a simplified visual stepping-stone stylus maze, and a simplified tactual maze, see Milner et al.[R18] However, for the incomplete pictures test, in particular, one can

argue that this is a very different learning task for normal subjects, who can rely on verbal memory as a guide, than it is for an amnesic patient. Once the normal subject has identified a picture, the name of the object is facilitated, so that when the most sketchy versions are presented a

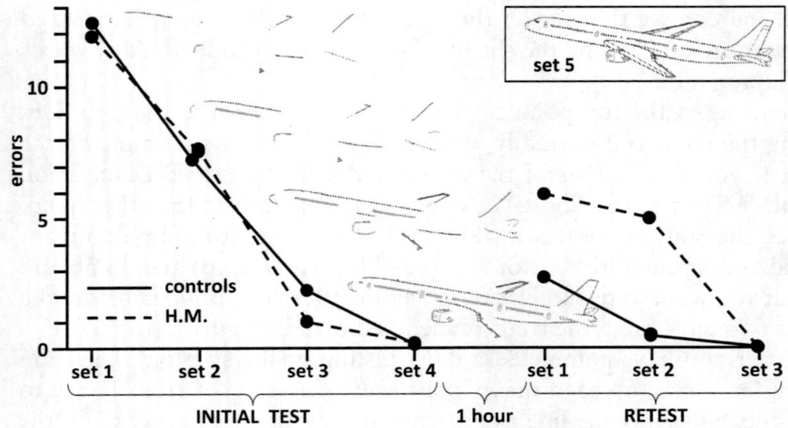

Figure 7.5 Perceptual memory in HM. The graph shows the results when testing HM on the incomplete pictures test (see text). The example images are a modernized version of one of the original test figures, see Milner et al. [R13, Fig. 11] On retest after 1 hour, H.M. had 50% fewer errors on images from Set 1 (the most incomplete pictures) but less improvement than controls matched for initial test performance. Both H.M. and the controls reached 0 errors on set 3 in the retest compared to Set 4 in the initial test. Data from Milner et al.[R13, Fig. 12] 'Set 5' example image by unidentified NASA employee – Dryden Aircraft Graphics Collection, Public Domain, via https://commons.wikimedia.org/w/index.php?curid=78975290.

> second time, [they have] many of the correct names ready to try out. Since the items of Set 1 are not similar to each other, normal subjects will be aided in their guessing by the limited choice of verbal labels [i.e., reduced interference]. We do not know if they would fare any better than H.M. without this added prop.
>
> (Milner et al.[R13, p. 231])

Interestingly, addition of cues also shows residual anterograde memory even for famous people.

> We used cued recall and forced-choice recognition tasks to investigate whether the patient H.M. had acquired knowledge of people who became famous after the onset of his amnesia. Results revealed that, with first names provided as cues, he was able to recall the corresponding famous last name for 12 of 35 postoperatively famous personalities. This number nearly doubled when semantic cues were added, suggesting that his knowledge of the names was not limited to perceptual information, but was incorporated in a semantic network capable of supporting explicit recall. In forced-choice recognition, H.M. discriminated 87%

> of postmorbid famous names from foils. Critically, he was able to provide uniquely identifying semantic facts for one-third of these recognized names, describing John Glenn, for example, as 'the first rocketeer' and Lee Harvey Oswald as a man who 'assassinated the president.' Although H.M.'s semantic learning was clearly impaired, the results provide robust, unambiguous evidence that some new semantic learning can be supported by structures beyond the hippocampus proper.
>
> (O'Kane et al.[R15, p. 417])

Consolidating the engram – "How do memories leave their mark?"[R3]

I hope I have, by now, convinced you that Hebb was largely right in seeing memories as stored in much the same circuits as register initial perception. Short-term (active/working) memory involves iterative refreshment of the relevant cell assemblies via circuits involving areas like frontal cortex. Ongoing activity in such circuits also allows cells to wire together when food or shock arrives even when warning stimuli like bells or tones have been absent for some time. This also appears to be the basis for H.M.'s, albeit limited, memories of very simple maze problems.

But, now, we need to ask how memories get maintained over long periods of time. How is it, for example, that Henry Mollaison's early memories remain intact but not memories of events after his operation? Or, rather, how is it that the latter seem to be retrievable but only with clues or other procedures that reduce the number of possible alternative answers? The fast answer is that initial learning becomes 'consolidated' over time. But what does that mean?

The slow answer requires a deeper look at the mechanisms of LTP and the involvement of sleep in consolidation and reconsolidation. But as is typical of biopsychology we must first ask: why do we sleep? Indeed, what is sleep? These are the questions of the next chapter.

Notes

1 For an interesting comparison of the parallel ideas of Hebb and Lashley, see Nadel and Maurer.[R14]

2 Decode for the Famous Faces Test example. This is not a properly constructed or normed version of the test but should give you an idea of what it feels like and, despite only N = 3 faces per major time period may give you the kind of memory curve that you will see in the figure showing H.M.'s results. Note that the faces are organized by time left to right and then top to bottom – they would be mixed up for a real test.

Hubert Humphrey	**Spiro Agnew**	**Walter Mondale**
Dan Quayle	**Al Gore***	**Dick Cheney**
Mike Pence	**Kamala Harris**	**James Vance**

*Note that Al Gore is likely to be remembered for his recent climate activism and so is not a particularly good example for this test.

References

R1. Annese, J, Schenker-Ahmed, NM, Bartsch, H, Maechler, P, Sheh, C, Thomas, N, Kayano, J, Ghatan, A, Bresler, N, Frosch, MP, Klaming, R, & Corkin, S, Postmortem examination of patient H.M.'s brain based on histological sectioning and digital 3D reconstruction. *Nature Communications*, 2014. 5: A3122.

R2. Baddeley, A, Working memory: Theories, models, and controversies. *Annual Review of Psychology*, 2011. **63**: 1–29.

R3. Bear, MF, How do memories leave their mark? *Nature*, 1997. **385**: 481–2.

R4. Corkin, S, What's new with the amnesic patient H.M.? *Nature Reviews Neuroscience*, 2002. **3**: 153–60.

R5. Fuster, JM, *Memory in the cerebral cortex*. 1995, Cambridge: MIT Press.

R6. Goldman-Rakic, PS, Working memory and the mind. *Scientific American*, 1992. **267**: 73–9.

R7. Hebb, DO, *The organization of behavior: A neuropsychological theory*. 1949, New York: Wiley-Interscience.

R8. Marslen-Wilson, WD & Teuber, H-L, Memory for remote events in anterograde amnesia: Recognition of public figures from newsphotographs. *Neuropsychologia*, 1975. **13**: 353–64.

R9. McNaughton, N & Gray, JA, *The neuropsychology of anxiety: An enquiry into the functions of the septo-hippocampal system* (3rd ed.). 2024, Oxford: Oxford University Press. https://global.oup.com/academic/product/the-neuropsychology-of-anxiety-9780198843313?cc=es&lang=en&.

R10. McNaughton, N & Wickens, J, Hebb, Pandemonium and catastrophic hypermnesia: The hippocampus as a suppressor of inappropriate associations. *Cortex*, 2003. **39**: 1139–63.

R11. Milner, B, Les troubles de la mémoire accompagnant des 1ésions hippocampiques bilatérales, in Milner, B. (ed.) *Physiologie de l'Hippocampe*. 1962, Paris: Centre National de la Recherche Scientifique. p. 252–72.

R12. Milner, B, Disorders of learning and memory after temporal lobe lesions in man. *Neurosurgery*, 1972. **19**: 421–46.

R13. Milner, B, Corkin, S, & Teuber, HL, Further analysis of the hippocampal amnesic syndrome: 14-Year follow-up study of H.M. *Neuropsychologia*, 1968. **6**: 215–34.

R14. Nadel, L & Maurer, AP, Recalling Lashley and reconsolidating Hebb. *Hippocampus*, 2020. **30**: 776–93.

R15. O'Kane, G, Kensinger, EA, & Corkin, S, Evidence for semantic learning in profound amnesia: An investigation with patient H.M. *Hippocampus*, 2004. **14**: 417–25.

R16. Ogden, J & Corkin, S, Memories of H.M., in *Memory mechanisms: A tribute to G.V. Goddard*, WC Abraham, M Corballis, et al., Editors. 1991, Lawrence Erlbaum. New Jersey p. 195–215.

R17. Pinel, JPJ, *Biopsychology* (4th ed.). 1999, Boston: Allyn and Bacon.

R18. Scoville, WB & Milner, B, Loss of recent memory after bilateral hippocampal lesion. *Journal of Neurology, Neurosurgery, and Psychiatry*, 1957. **20**: 11–21.

R19. Stephan, H & Andy, OJ, Quantitative comparative neuroanatomy of primates: An attempt at a phylogenetic interpretation. *Annals of the New York Academy of Sciences*, 1969. **167**: 370–87.

R20. Strange, BA, Witter, MP, Lein, ES, & Moser, EI, Functional organization of the hippocampal longitudinal axis. *Nature Reviews Neuroscience*, 2014. **15**: 655–69.

8 Why do we sleep?

What is sleep?

Like 'What is an emotion?', 'Why do we sleep?' seeks a simple answer. But like an emotion, your sleep has a range of unrelated parts. Unlike an emotion, sleep's parts can fulfil quite different functions. Sleep is just when they choose to happen.

You can sometimes 'go to sleep' during the day. But what we normally call 'sleep' typically takes up 6–10 hours of our night. Reptiles show a similar pattern of 'adaptive inactivity' – where inaction keeps them out of trouble. But their sleep does not appear to have the complex structure of ours.[R28] For example, they appear to lack Rapid Eye Movement (REM) sleep and so do not dream.

Our night-time sleep has structure. In particular, there are regular periods of REM sleep. During REM, someone's eyes can be seen moving under their eyelids if you look carefully. If you are woken during REM, you will usually report a dream (so sometimes it's called dream sleep). Non-REM sleep can also occasionally contain dreams or nightmares and, importantly, involves a different brain state from REM.

All mammals (including the egg-laying monotremes like the platypus) sleep. This sleep has a similar complex structure to ours. But the amount of REM sleep (essentially disconnected from the world) is less in species that are more subject to predation, and more in less predated species.[R2] Also, total sleep time appears to vary. "Lions sleep long and deeply, whereas one example of their prey, giraffes, have one of the lowest recorded sleep durations and must not sleep deeply if they are to survive".[R28, p. 747] The fact that giraffes sleep at all suggests a direct need for at least some sleep processes. Notably, marine mammals, particularly dolphins that swim continuously throughout their lives, cannot sleep like other mammals without drowning. Nonetheless, they show non-REM sleep. But this occurs only on one side of the brain at a time and then alternates to the other. The transition that occurred in their evolution must have been similar to the pattern for fur seals, which show unilateral non-REM sleep only when in the water (see section on evolution below).

DOI: 10.4324/9781003465027-8

It looks like evolution has taken the opportunity to use the available down-time of sleep for a Smörgåsbord of processes that would not fit so well with the brain states we normally have when we are awake and where some of these require sleep. The function of such processes (e.g., release of growth hormone) need have no link to the function of the bodily inactivity (to keep you safe) – and so need have no simple relation to total sleep time. It is also clear that the functions of REM and non-REM sleep are at least somewhat distinct given their different cross-species correlations.

That said, I will argue below that *in humans* non-REM and REM sleep make complementary contributions to consolidation of various types of memory. So, following on from Chapter 7, we will look at consolidation both in sleeping and in waking. Other sleep processes appear important for brain health, protecting against neurodegeneration,[R19] and also for mental health.[R8] With some mental disorders, disturbed sleep is both a cause and a symptom (for example Riemann et al).[R23] For less extreme examples of the impact of sleep disturbance on mental functioning, just observe parents with their first[1] new-born child; or note the severe jet-lag-like effects "when social timing forces an abnormal pattern of sleep and wake upon our endogenous sleep biology".[R8, p. 326]

Our night-time sleep is locked to an internal clock. This has a free-running time under fixed lighting conditions that for most of us is a bit longer than 24 hours. It is reset to some extent each morning by bright sunlight to keep it in synchrony with the light-dark cycle in the world.

This clock is why we can suffer from jet lag (it tries to keep us to 'home' time), and why recovery from jet lag is helped by exposure to early morning sun (pushing the reset mechanism). Regular inactivity (whether diurnal sleep or annual hibernation) seems to have evolved to allow animals "to *anticipate* periods that are not optimal for survival and propagation".[R28, p. 747, added emphasis]

The complexity of the question 'What is sleep?' should now be a bit clearer. So, let's look at sleep's parts and its evolution in more detail. This should provide answers to the question of 'Why do we sleep?', especially in the sense of 'What is it for?', with different answers for different types of sleep, all enabled by the original global function of sleep: the adaptive control of motor activity.[R27,28]

Stages and types of sleep

The idea of different stages of sleep and its division into two major types is based on non-invasive recording of scalp EEG in healthy volunteers. As you relax and drift off to sleep, your brain produces bursts of alpha waves (α, 8–12 Hz[2]). These α bursts get progressively fewer and disappear as your sleep gets deeper. Then, more and more delta waves (δ, 0.5–4 Hz) start to appear and progressively increase in number. The deepest sleep is almost all δ and is usually called Slow Wave Sleep (SWS).

This first period of sleep is all non-REM sleep and is conventionally labelled as having four stages (with SWS being Stage 3 and Stage 4). But, as you will see if you check out Experiment 8.1, there are no clear divisions between these stages. To make this point, in the figure in the experiment, I have put the label S2.5. No one normally does this (they would say 'S2/S3'), but it would make sense as the variation between stages is quantitative rather than qualitative, as the experiment demonstrates.

After a period of SWS, your EEG rapidly changes to ~Stage 2 and then to a pattern that looks very like the waking one. But you are not only still asleep, but you are also paralyzed from the neck down! While your neck muscles go silent (see the bottom trace in the figure in Experiment 8.1), your eyes start darting rapidly around under your closed eyelids – you are in REM sleep. Having a waking pattern EEG is consistent with your (usually unremembered) experience during dreaming. You dream regularly (about every 60–90 minutes) *throughout every night*. But even if you wake from a dream, your conscious memory of it fades very fast. We will come back to this when considering consolidation of memories.

Experiment 8.1 Seeing brain activity change

The main part of this experiment is observational. I'll suggest some ways you can record your own EEG at the end. The figure gives you samples of EEG (all obtained from a single person), annotated with the conventional sleep stages from an older classification system that has now been superseded. But it's easy to understand; and, if you want them, there are a lot more examples of EEG and commentary on it at https://archive.org/details/RKManual/mode/2up?view=theater.

1 Check out the top trace, Stage W (waking). This is during relaxation before sleep. The upper channel is muscle activity (EMG); the lower one is scalp (EEG). The key thing to note is that the EEG is generally even with a lot of fast activity (almost looking like fuzz) and occasional spikes.
2 Now look at the second trace. This shows the transition from W to Stage 1 (W→S1). Bursts of α (8–12 Hz) are flagged with a dashed line. The α you see here during waking is often used by biofeedback systems that try and train people to relax. Note whether you can see a clear transition between W and S1.
3 Look at the third trace (S1). Note whether you can see any α.[3]
4 Look at the fourth trace (S2). Note the sharp waves.
5 To get your eye tuned, now look at the S4 trace. Note the size and shape of the δ waves.
6 Look at S3. Note the substantial number of δ waves.

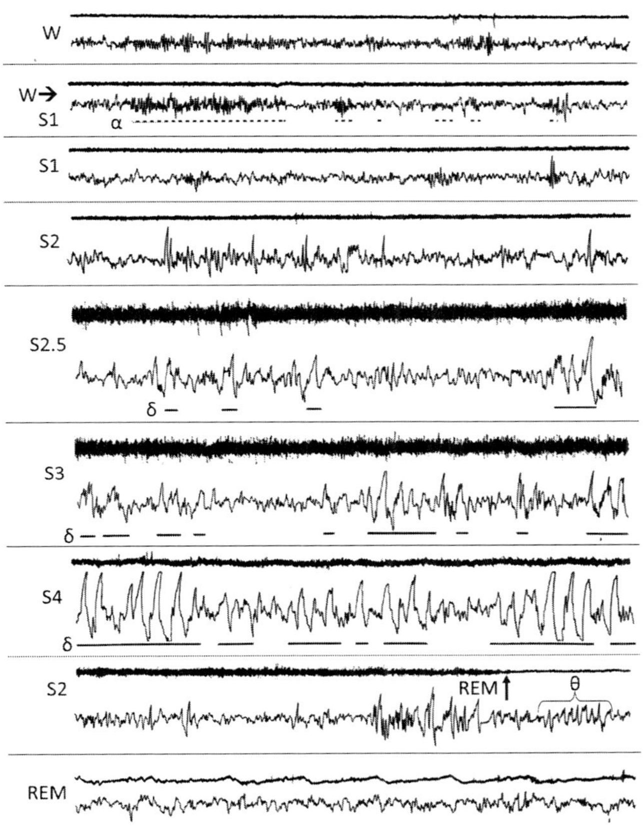

Images from[R22] obtained via Public Domain. https://archive.org/details/RKManual/mode/2up?view=theater

7 Now look at the trace I have labelled S2.5. Ask, first, can you see δ waves (like Stage 3)? Next, can you see sharp waves (like Stage 2)? Hopefully, you now see why I labelled it 2.5. But this label is just a joke; the original is classified as "almost borderline between Stages 2 and 3". No one would call it 2.5 – but perhaps they should?

8 *You are now ready for the observational experiment.* First, make a copy of the EEG and print it. Next, go over the traces and make sure you agree with the underlining marking the periods with δ waves. Similarly mark any that you think they missed. Now, with a ruler, measure the length of each underline and write the value above it. For each of S1, S2, S2.5, S3, and S4, separately, add the lengths together and divide by the total length of the trace and multiply by 100 (to get a percentage). Plot a graph of the result. How close is this to a straight line between S2 and S4? Is this consistent with the stages being points along a smooth increase of δ waves within a single type of sleep?

9 Now look at the bottom trace (pure REM). Note that the EEG looks very like waking, with no α, δ, or sharp waves. Also note that the EMG trace is not fuzzy (for an extreme comparison, compare S3).

Now look at the next to bottom trace. Its EEG starts off as S2 with strong sharp waves 2/3rds along the trace. The EMG fuzz is consistent with this. But at the end, the EMG has gone flat indicating REM. The EEG is consistent with this and includes a burst of θ (4–7 Hz) rhythm, which is typical of REM but also occurs during waking. Your observation question is: can you see a smoothly graded change (as you saw with δ waves)? Or, is there a sudden S2/REM transition? If the latter, is this evidence for two distinct types of sleep: non-REM and REM?

To do a real experiment using EEG in general and looking at sleep in particular would have been impossible at home until very recently. However, what is possible and the price ($150–$500) has become very interesting recently (try Googling 'wearable EEG'). Be careful about what you think you are buying. If you just want to check how to use EEG biofeedback for relaxation, there are many off-the-shelf systems. But they likely do not tell you *how* they work (by using α-based measures of relaxation, see W-S1) and may not allow transfer of the original EEG to your computer.

Some of the systems can be used for research but it can be fiddly and device-dependent to set up. This is why I have not given you instructions for doing this. A computer science friend of mine has used a commercial cheap headset to allow mental control of a drone (in a lecture theatre!). It takes a bit of practice to control it properly but it's pretty cool. But, for that you need to capture the EEG electronically in real time, process it correctly in real time (are you going to use α, θ, δ, γ, or what, power) and interface it with your drone control. If you are a computer geek, I hope I whetted your appetite. If not, you have been warned.

Even better is where such devices can be used for mental control of wheelchairs for people who are severely physically disabled. If you are in that situation, get one of the professional setups.

If you want to try sleep recording you will likely need to use a swimming cap or similar to hold the gear on (and make sure any cable will not strangle you). Make sure you allow for at least one night of abnormal sleep. Sleep laboratory work usually has a night or two of 'dummy' recording to allow the participant to adapt to the wiring and the laboratory. You likely don't sleep as well in a hotel room as you do at home.

The evolution of the types of sleep

Dormancy occurs even in organisms that do not have a nervous system. Species comparison suggests that sleep proper developed later, and non-REM and then REM types were added in succession (Figure 8.1).

What is, behaviourally, diurnal sleep occurs in reptiles; but with no Stage 4 or REM; birds have Stage 4 with a small amount of REM; Platypus (an egg-laying mammal) have extensive Stage 4 and substantial REM, restricted to the subcortex; marsupials and other mammals have cycles of SWS and REM (with substantial Stage 4 in the initial cycles). Dolphins, as

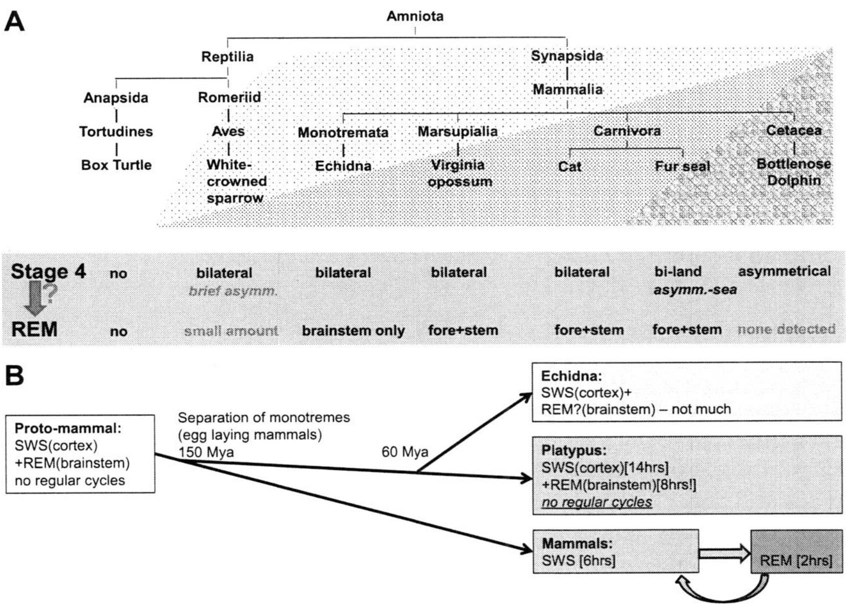

Figure 8.1 Evolution of sleep. (A) A phylogenetic tree for tetrapods with specimen examples that have been assessed for SWS and REM sleep. Shading shows progression from no SWS or REM, to SWS without REM, to SWS with REM, and finally to asymmetrical SWS (where there may also be an asymmetrical REM state but distinguishable from waking). The monotremes are an intermediate case where there is REM but only in the brain stem not cortex. Fur seals are an intermediate case with standard SWS + REM on land but usually asymmetric SWS in the water. Figure based on Siegel.[R28] (B) Assumed evolution from the proto-mammal common ancestor. Monotremes split off from other mammals about 150 million years ago (Mya) and Echidna from platypus about 60 Mya. Echidnas have cortical SWS but only marginal REM, restricted to the subcortex. They are taken here to be similar to the proto-mammal. Platypus added more subcortical REM but without regular cycles. Other mammals must have increased their REM independently (since they split off from monotremes before echidna) and added regular SWS ⇔ REM cycles. Note that with later cycles in the night SWS amount decreases, and it does not usually reach Stage 4.

noted earlier, have Stage 4 that alternates from one side of the brain to the other. As an indication of how this could have evolved, fur seals are like other mammals on land but have substantial dolphin-like asymmetry in the water.

> On land, sleep in the fur seal generally resembles that in most land mammals. The EEG is bilaterally synchronized and the animal closes both eyes, seems unresponsive and cycles between REM and non-REM sleep. By contrast, when the fur seal is in the water it usually shows an asymmetrical pattern of behaviour, with one of the flippers being active in maintaining body position while the other flipper is inactive.
>
> (Siegel[R27, pp.749–50])

This progression fits with the idea that sleep, per se, has adaptive value. Then, once it was established, it enabled the evolution of SWS sleep, which did not immediately require REM to fulfil an additional adaptive function. REM sleep was then added, followed by cycling between SWS and REM. This cycling suggests that, at least in part and in some species, the functions of SWS and REM interact – and I will argue that this interaction can include refined consolidation and re-consolidation of memories.

Is sleep necessary? The case of Randy Gardner

This evolutionary story strongly suggests that the different types of sleep have important different functions. If all sleep does is keep you out of trouble, then being kept awake should not be problematic. Even if it involves rest and recuperation, you should be able to do this (as do reptiles) without having SWS and REM.

On an analogy with eating or drinking, if you are sleep-deprived *and* if sleep is important, you would expect rebound increases in sleep when you are allowed to sleep again to make up the loss. If you regularly use an alarm clock to wake during the week, you likely find that you want to 'sleep in' during the weekend to catch up. You would also expect to see adverse changes while you are being deprived. I already talked about social jet-lag and the problems for first-time parents. What about more direct attempts to see the effects of sleep deprivation – particularly any changes in non-REM and REM sleep?

A much-quoted case is Randy Gardner. (For more of his story, see https://en.wikipedia.org/wiki/Randy_Gardner_sleep_deprivation_experiment, which also comments on other similar record attempts.) He was a 17-year-old who wanted to break the 'staying awake' record. He managed 11 days (264.4 hours). It is not usually mentioned that he had previously gone for 252 hours without sleep as a kind of 'warm up'. He was monitored by a range of people, including a prominent sleep researcher (Dr William Dement) and a neurologist (Lieutenant-Commander John Ross) over the last 3 days – with three neurological examinations during that time and three more (at 10 days,

6 weeks, and 7 months) after. My version of the results is based on the original reports and differs somewhat from that usually found in the textbooks.

In terms of rebound, after the ~15-hour first night, Randy returned to a standard 8-hour pattern. But what is of interest is the amounts of types of sleep obtained. When you stop dieting you usually regain (or exceed!) your previous body weight. What about sleep?

A really exciting aspect of this story is that Randy had EEG recordings made (hence William Dement's involvement). Randy had done an unofficial trial run and so could get professionals (science and psychiatry and neurology) involved for his San Diego Science Fair project (he was only 17).

My take on the EEG data is not the standard story.[4] Figure 8.2 uses the original reported data,[R10, p. 32] so that you can make up your own mind. You can see that Randy's SWS (i.e., S3 + S4) and REM sleep recovered steadily, with the same amounts, and at much the same rate as each other over the first 3 days. If we include an estimate for day 4 using the regression lines in the figure, he recovered at least 90% of both SWS and REM over 4 days. Given that total sleep time recovered by day 2, including both rebound SWS and REM was at the expense of Stages 1 and 2. So, SWS and REM appear more important than S1 + S2. The parallel values are interesting. This suggests that in humans, the two may have become linked functionally. Whereas, we saw that across a range of species, they varied from each other. I will come back to this when discussing memory consolidation.

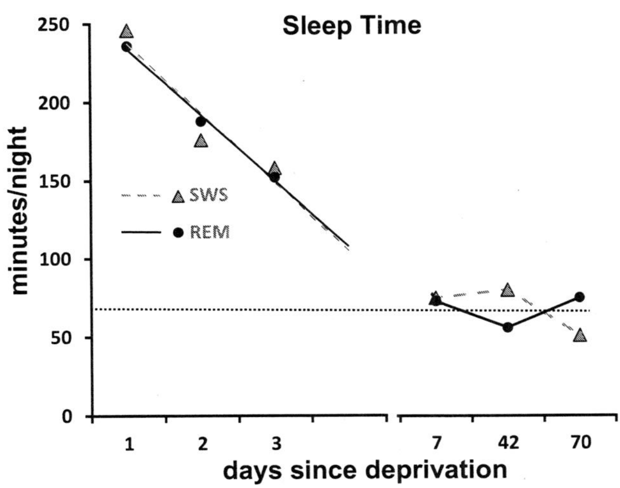

Figure 8.2 Randy Gardner's sleep during recovery. This shows his SWS and REM sleep times after deprivation data from Gulevich et al.[R10, p. 32] with regression lines extrapolated to day 4. Based on the average of days 7, 42, and 70 (dashed line), he recovered 91% of his lost SWS and REM over the first 4 days.

Randy is often reported as having had no problems during his deprivation or afterwards. But here is part of the neurologist's report.[5] This includes corroboration from other cases with less extreme deprivation, and so, Randy is not a special N = 1 case.

> During the **second day** of the vigil the patient noticed difficulty focusing his eyes. After the second day he stopped watching television for the rest of the experiment because his eyes were so extremely heavy and tired. At this time he also had mild astereognosis[6]. By the **third day** he developed mood changes, ataxia, difficulty saying tongue twisters, and marked nausea. During the **fourth day** he became irritable and un-cooperative; he developed memory lapses, difficulty concentrating, the feeling of a tight band around his head, and saw fog around the street lights. About 0300 that night he experienced the illusion that a street sign was a person. A short time later he imagined he was a great Negro football player and resented statements made about his ability and the Negro race. By the **fifth day** his equilibrium was normal, but he had intermittent hypnogogic reveries, such as seeing a path running through a quiet forest and plants in a garden. On the **sixth day** ataxia, mild astereognosis, and dysnomia were again present, and during the **next three days** slight irritability, mild uncooperativeness, memory lapses, and inability to concentrate reappeared. Intermittent slurred speech was evident on the seventh day and it continued to deteriorate to a soft, slow, slurred, mushy quality by the end of the wakeful period. By the **ninth day** he occasionally thought in fragments and frequently did not finish sentences. Also, blurred vision became more of a problem. During the last two days of the vigil he felt a certain radio entertainer on whose show he later appeared was trying to make him appear foolish because he could not recall certain specific facts about the vigil. The patient finally went to sleep after 264 hours of wakefulness (Jan 8, 1964) and slept for 14¾ hours. The reason for terminating the vigil was simply that the patient had achieved his goal.
>
> (Ross[R24, pp.399–400, emphasis added])

This likely fits with the impression you have of the partially sleep-deprived parents of your acquaintance. It is far from 'no problems'. I would argue, not just from this, that if you do use an alarm clock, you should find ways of not needing it. A good guide is that if you sleep longer when you don't use the alarm, your normal routine is generating cumulative deprivation and messing up your intellectual and emotional performance during the day at least a bit. If you don't need to sleep in, you should be waking naturally at a fixed time, so why are you using the alarm? We will come to other aspects of sleep hygiene later.

In relation to what I will say about consolidation of memory, note that the problems that Randy experienced involve mood, cognition, perception, active memory, attention, even sanity. Indeed, there is a history of sleep deprivation being reported as producing a kind of psychosis.[R10] So, as we already concluded, sleep types involve a lot of functionally different stuff. Randy was not tested for long-term memory of any kind, and so does not tell us anything about consolidation.

Are non-REM and REM sleep equally necessary?

We may seem to have already answered this question via Figure 8.2. But Randy (and equivalent animal experiments) had total sleep deprivation. Maybe the driver is just SWS and REM came along for the ride, or vice versa? We cannot selectively deprive mammals of *all* their SWS. They have to go through SWS to get to REM. But you can selectively nearly totally deprive mammals of REM by waking them up as soon as their rapid eye movements start.

Unfortunately, the techniques used to wake up rats have problematic stress-related effects; other animals give different results, and pushing humans to the limit has ethical problems. However, partial deprivation can be more balanced during the first 5 hours of sleep. Selective partial REM deprivation causes a REM but not non-REM rebound, whereas equivalent amounts of selective partial non-REM deprivation cause non-REM (including SWS frequencies) but not REM rebound.[R5] This makes both look functionally important.

But is part of their functional importance linked to consolidation of memories or something else?

> It has been hypothesized that REM (rapid eye movement) sleep has an important role in memory consolidation. The evidence for this hypothesis is reviewed and found to be weak and contradictory. Animal studies correlating changes in REM sleep parameters with learning have produced inconsistent results and are confounded by stress effects. Humans with pharmacological and brain lesion-induced suppression of REM sleep do not show memory deficits, and other human sleep-learning studies have not produced consistent results. The time spent in REM sleep is not correlated with learning ability across humans, nor is there a positive relation between REM sleep time or intensity and encephalization across species. Although sleep is clearly important for optimum acquisition and performance of learned tasks, a major role in memory consolidation is unproven.
>
> (Siegel[R26, p. 1058])

I speculate that different types of memory depend on sleep differently; that declarative memory can be consolidated during waking; and, also, declarative consolidation (and reconsolidation) may depend on sleep only for longer retention intervals. That is, consolidation is easier to produce during sleep than during waking. In particular, re-consolidation during sleep may occur at longer durations after learning than deprivation experiments can technically achieve. Critically, sleep processes, particularly with REM, during sleep can also occur when awake. So, as with the takeover of S1 + S2 by S3 + S4 and REM after deprivation, some processes normally occurring in SWS and REM can take over awake down-time.

An additional issue is that sleep seems more important for emotional habituation and procedural memory than declarative (event-related) memory. "In contrast to the declarative system, evidence for the reliance of procedural memory on sleep in humans has been incredibly robust, and currently offers the most promising and informative model of sleep-dependent learning."[R31] As we saw in Chapter 7, these different forms of memory involve different learning processes.

On the face of it, given the evidence we have looked at so far, sleep is not necessary for consolidation of simple purely associative memory ("cells that fire together wire together") but is necessary for reinforcement-based memory and other processes like the habituation of anxiety that are linked to motivation. So, let's take a close look at how such memory is consolidated and re-consolidated in the longer term and try and answer the question we asked at the end of Chapter 7. "How do memories leave their mark?"[R4]

LTP – 1, 2, 3

Chapter 7 presents simple association as depending on Long Term Potentiation (LTP) and dependent on activation of NMDA receptors. But that is just one form of LTP that I will refer to as LTP1. LTP1 is relatively short-lasting (Figure 8.3A). The longer-lasting LTP2 requires co-operation of the NMDA receptor with an additional receptor. The even longer-lasting LTP3 does not involve the NMDA receptor. LTP2 and LTP3 involve protein synthesis via RNA translation and LTP3 seems more dependent than LTP2 on DNA to RNA transcription. The protein synthesis and transcription are the basis for synaptic remodelling – the ultimate 'wiring together' process. For more details of these types of LTP, the receptors involved, the ways they can be selectively blocked and the biochemical signalling cascades that they trigger, see Raymond.[R21]

But all of these types of LTP normally show decay that appears to return to pre-stimulation values (Figure 8.3A). That said, LTP has been shown to reach an asymptote 10% above the original baseline values (i.e., becoming essentially permanent); and can last at least 360 days (about 1/3rd of the rat's life span; Figure 8.3B). Since the rat was allowed to sleep during this time, at least part of this very long maintenance of potentiation could have been linked to SWS and/or REM sleep.

Walker's procedural reprocessing theory

As we have seen, procedural memory is more obviously dependent on sleep than is declarative memory. But even procedural memory has aspects of its

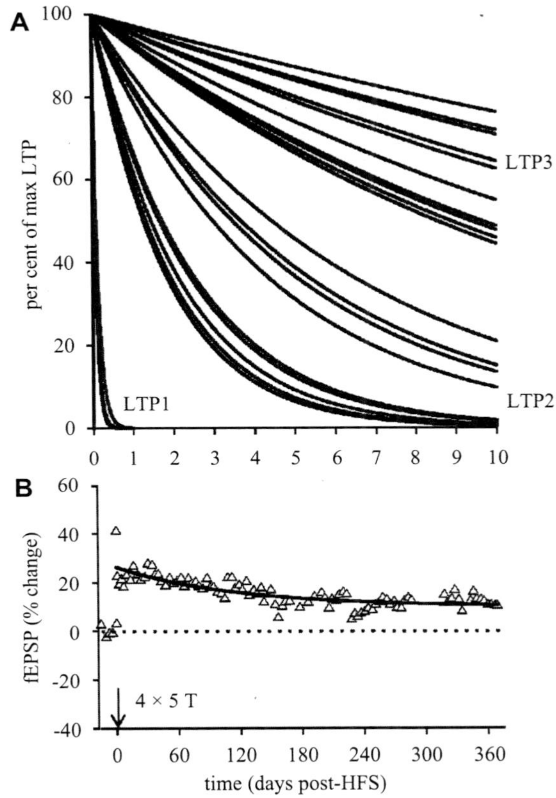

Figure 8.3 Time course of different types of LTP. (A) Decay curves are exponential and can be divided into distinct LTP1, LTP2, and LTP3 groups depending on their time constants (2.1 hours, 3.5 days, and 20.3 days – with this value, LTP3 will decay to below 5% of its original amount in less than 100 days and appear to return to baseline. (B) Data from a single rat given four bouts of five trains of stimulation. The fitted exponential has an asymptote of 10% and lasts at least a year. Figures from Abraham,[R1] with permission of The Royal Society (U.K.); permission conveyed through Copyright Clearance Center, Inc.

consolidation that do not require sleep. Matthew Walker's[R31] "refined model of sleep and the time course of memory formation"

> offer[s] a new neurocognitive framework of *procedural learning*, consisting first of acquisition, followed by two specific stages of consolidation, one involving a process of stabilization, the other involving enhancement, whereby delayed learning occurs. Psychophysiological evidence indicates that initial acquisition does not rely fundamentally on sleep. This also appears to be true for the stabilization phase of consolidation,

> with durable representations, resistant to interference, clearly developing in a successful manner during time awake (or just time, per se). In contrast, the consolidation stage, resulting in additional/*enhanced learning in the absence of further rehearsal*, does appear to rely on the process of sleep, with evidence for specific sleep-stage dependencies across the procedural domain.
>
> (Walker[R31, p. 51, added emphasis])

The stabilization phase consolidates the initial memory representation into an interference-resistant form, with no change in the level of performance. Enhancement occurs only at the next available period of sleep. "NREM and REM sleep reactivation ('replay') of local networks established during waking continues to facilitate coincident firing between pre- and postsynaptic terminals during sleep".[R31, p. 59] This replay wires cells together and results in improved performance similar to that which would be produced by additional training.

Walker's picture of how NREM and REM interact is not quite the same as the one I will present for declarative memory shortly. But procedural and declarative learning are not the same. Procedural learning requires reinforcement (not association), involves multiple trials to achieve (rather than a single episode of learning), and cannot easily be replayed "offline" during the wake state (as opposed to being strengthened with additional learning trials).

SWS and declarative reconsolidation

A background fact of interest is that growth hormone is released at the same time as the first episode of SWS (Stage 3 + Stage 4). The two are not directly linked, but it is clear that the hormone could enable synaptic growth or change.

One clue to what happens in SWS is replay of cells' firing patterns. Figure 8.4 shows part of the results of a cunning experiment carried out by William Skaggs and Bruce McNaughton.[7] They recorded from rats during sleep before and after running on a triangular track. During running, hippocampal cells showed the typical 'place fields'[8] – firing consistently in a particular location (the middle of the left-hand arm, in this case). Importantly, pairs of cells showed a consistent relationship in their timing, based on their place fields being at different positions that were always in the same order because the rats always ran around the track counterclockwise (Figure 8.4). During SWS, after maze running the ordered linkage of the cells matched what had been observed during running – appearing to replay the waking experience. Importantly, ordering was negligible before training and so this replay involved new connections and could potentially involve their re-processing during sleep.

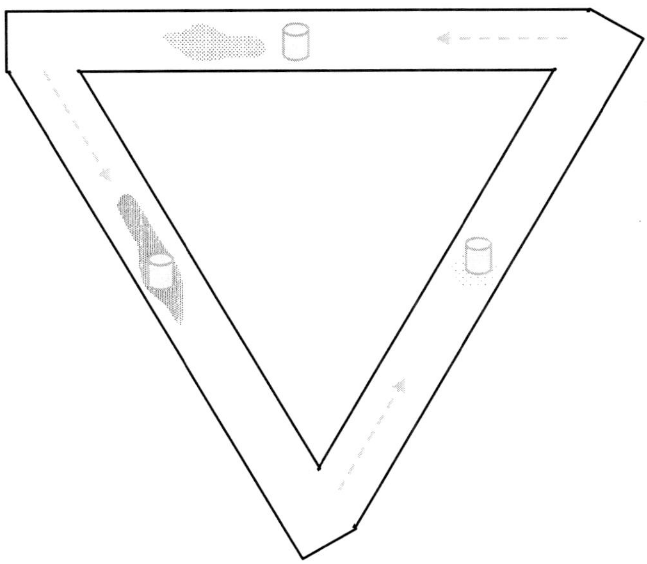

Figure 8.4 Place cell firing on a triangular track. Diagram based on data from Skaggs and McNaughton.[R29] A rat ran counterclockwise on a track for small food pellets placed in the middle of each arm. Cell 1 fired most in one place (the middle of the left arm). Cell 2 (not shown) was similar but its place was slightly behind Cell 1.

This replay is likely to have enhanced the memory.

> Using artificial patterns of olfactory bulb stimulation in a fear conditioning procedure in the rat, we tested the effects of imposed post-training replay during SWS and waking on strength and precision of pattern memory. … Imposed replay during post-training SWS enhanced the subsequent strength of memory, whereas the identical replay during waking [but without the fear reinforcer] induced extinction. … SWS is ideal for replay given hyporesponsive sensory systems, and thus reduced interference. … Sleep-dependent consolidation has been linked to emotional, procedural, and declarative memory in human and nonhuman animals.
> (Barnes and Wilson[R3, p. 5134])

You might think, then, that SWS simply replays memories to strengthen them further through a simple "cells that fire together wire together" mechanism. There are two reasons to see what is going on as not simply strengthening but rather quality enhancement (as Walker emphasizes) and some reduction in strength. One reason is the cyclical following of SWS by REM. The second is the decrease in SWS over sleep cycles that appears to be linked to synaptic 'pruning'.

To sleep, perchance to prune[R30]

Giulio Tononi and Chiara Cirelli argue that the delta waves of SWS at the start of sleep are strong because the synapses involved in replay were strengthened during wake. Sleep processes weaken these synapses, conserving energy, and preventing memory stores from becoming oversaturated. That is why delta waves progressively weaken over the sleep cycles. "Spontaneous firing selectively removes or weakens neuronal links. *Somehow unimportant links get weakened more than significant ones do*, allowing important memories to remain intact.",[R30, emphasis added] One possibility is that LTP is strengthening stronger memory traces (as with the rat replay experiment) while LTD is weakening weaker ones.[R31]

> Memory consolidation benefits from sleep. In addition to strengthening some memory traces, another crucial, albeit overlooked, function of memory is to erase irrelevant information.... Our data indicate that relevant items that subjects strived to memorize are consolidated during sleep to a greater extent than items that participants did not intend to learn. This process appears to depend on a differential activation of the hippocampus at encoding, which acts as a signal for the offline reprocessing of relevant memories during postlearning sleep episodes.
>
> (Rauchs et al.[R20, p. 2563])

One important feature of replay (whether during sleep or wake) is that activation of a memory appears to make it available for reconsolidation or elimination ("cue-dependent amnesia"[R25, p. 1]) by manipulations such as Electroconvulsive Shock Therapy (ECT) or strengthening by, e.g., stimulation of brain arousal systems.[R25] Most easily demonstrated in rats, it can also be seen in humans.

For example, Maximilian Gahr and colleagues report a case[R9, p.E38] of treatment-resistant PTSD with traumatic memories of both a serious traffic accident and childhood sexual abuse. Just prior to ECT, "the patient was advised to remember the [traffic] accident and to describe his particular memories verbally. ... PTSD-symptoms related to the accident subsided completely and though [other] PTSD-symptoms persisted; they were not accompanied by memories related to the accident".

So, replay in SWS could strengthen a specific memory (through LTP of the cells firing together). But it could also weaken competing memories. This adaptive increase in memory contrast is sufficient to account for the evolution of SWS, independent of REM.

However, replay in SWS could also make a trace more likely to change its conformation later, during REM, than if recall had not occurred. We will see that increased and decreased synaptic strength can be concurrently advantageous even for a single engram. This is controlled by processes that occur most clearly in REM sleep.

With both within-item and item-wise changes, a key result will be the reduction in the amount of memory-space required – an increase in the efficiency of engram storage. Synaptic elimination and weakening will reduce the chance of saturation of the system, which would produce a resultant collapse of memory. So, memory improvement in sleep would be a non-specific system-wide result as much as a trace-specific one. This non-specificity could be one reason that it has been hard to demonstrate declarative memory problems with sleep deprivation.

REM, declarative reconsolidation, and theta rhythm[9]

SWS can easily be passed over as 'just sleep' – until you look at its fine-grain details. But REM seems truly mysterious. It is when we dream – which gave Hamlet pause. It arrived evolutionarily late, and so is not just a simple extension of SWS. (Neither is SWS an extension of 'simple sleep'/torpor.) In the mammal line, REM spread from subcortex, where it occurred most of the time, to later include cortex and started occurring in tightly packaged cycles.

REM is often called paradoxical sleep. It is when your EEG looks most awake, but your body looks most asleep – paralysed from the neck down.[10] Absent that paralysis, you would act out your dreams. This happens in a rare condition, REM sleep behaviour disorder.[R13] Likewise, cats given dorsal brain-stem lesions can appear to be hunting rats that are not there. Their "behaviour is comprised of violent phasic behaviour resembling predatory attack (pouncing with the forelimbs) interspersed with periods of quiet staring or searching movements".[R7, p. 234] REM sleep behaviour disorder is not 'sleep walking', which is an SWS state and guided by external input.

It's remarkably easy to tell if a rat is in dream sleep (REM). Record from its hippocampus and from its neck muscles. If the hippocampus shows continuous 'theta'[11] rhythm (also typical of when it is running around awake) and the neck muscles are silent then the rat is in REM. You could almost go just with the neck muscles, except they would be similarly silent if the rat was close to dead or had been anaesthetized. Yes, it's hard for us to believe that rats have the conscious experience of dreaming. But the brainstem-lesioned cats that we met earlier, hunting invisible rats while asleep, are hard to interpret any other way.

> Awake animals seemed to show theta rhythm when they were behaving in ways crucial to their survival... Predatory behaviour in the cat, prey behaviour in the rabbit, and exploration in the rat... (For example, a hungry rat will explore before it eats food even if food is placed in front of it.)... I suggested that [during hippocampal] theta rhythm... information essential to the survival of a species... was reprocessed into memory during REM sleep.
>
> (Winson[R32, p.44])

This reprocessing appears to involve enhancement rather than just strengthening of memory traces (echoing changes in SWS). Theta rhythm results from regular inhibitory input to the hippocampus that results in regular (~100–200 ms apart) periods of depolarization (increasing the chance of cell firing and so producing LTP) and hyperpolarization (decreasing the chance of cell firing and so producing LTD).[R32] This would interact with ongoing replay but not affect cell assemblies that are silent and be selective about strengthening versus suppression.

We already saw that processes in SWS can increase the relative strength of engrams of important compared to unimportant items. However, during wake, the hippocampus (and particularly its theta rhythm) appears important for suppressing equally strong but incorrect competing engrams.[R14] A similar action of theta rhythm during REM would produce a more subtle, content-guided, reconsolidation tidying up remaining problems not solved by SWS. Hence, the observed cycles of SWS are followed by REM, with more extensive REM in later cycles. Consistent with this, replay during SWS accelerates the drift from recently learned patterns to those seen in later memory recall; while REM counteracts the drift.[R6]

But, how does weakening memories improve them? With SWS the crude and simple answer was that you weaken unwanted (untagged) memories more than wanted (tagged on first creation for later off-line processing). Since both unwanted and wanted cell assemblies were strengthened while awake, the net result of sleep is weaker 'signal' memories but much weaker, potentially eliminated, 'noise' memories. 'Memory' is then better with a weaker signal because it has a hugely improved signal:noise ratio.[12] Oh, and the noise memories used up memory space unnecessarily, and the signal memories do not need as much space as they originally took up.

But with REM, the answer appears to be more complex and dynamic. One theory of what is happening here is that, during wake, episodes are encoded into hippocampal and shorter-term networks outside the hippocampus (with the hippocampus acting as a temporary store). During SWS repeated reactivation of the hippocampus (replay) transfers its part of the engram to a long-term store outside the hippocampus. In the process, the trace becomes more of a 'decontextualized' schema and is subject to downscaling that erases non-overlapping portions of concurrently re-activated cell assemblies.[R11, Figure 2, p. 84] I think this theory presents a good picture of declarative reconsolidation but with three gaps: (1) why is the hippocampus necessary only for memories subject to interference? (2) How does a single engram get transferred between three different memory stores? (3) What is the role of theta rhythmicity – which, as a simple timing mechanism, appears essential for encoding[R16] and is important for the bulk of hippocampal behavioural control.[R15]

If we take a Hebbian cell assembly view of the engram and add in REM (with its constant hippocampal 'theta' rhythmicity), I think the story can be simpler and more subtle. I have explained elsewhere[13] that the hippocampus operates to *suppress interfering* incorrect memories, *not produce* correct ones. In my view, the hippocampus prevents blocking of recall by competing incorrect items (think tip of the tongue phenomenon). This explains why interference-prone memories (often complex and declarative) are more dependent on the hippocampus. But the hippocampus does not make or store memories. Instead, at encoding, consolidation, and during reconsolidation (in REM sleep), it enhances memories, sculpting the activity of cell assemblies into a more optimal form.

At initial encoding, complex memories (including contextual information) will involve a range of separate cell assemblies and each of these is likely to involve many more connections than are strictly needed for later recall. The hippocampus, and particularly its theta rhythm, filters this pandemonium to improve the signal-to-noise ratio that is biased by content not just strength or previous tagging. I see the hippocampus as suppressing unwanted connections in other areas, not storing parts of the engram in its own circuits. Parallel distributed circuits run through the cortex then to the hippocampus and then back (via the thalamus) to re-cycle through the cortex.[R17] The engram is iteratively processed and adjusted (via LTP and LTD) at each cycle so long as its basic circuits are active. Information travelling round these long loop circuits is packaged by the theta rhythm to avoid mixing of different stages of the progressive enhancement of the engram. The same will occur during reactivation during sleep. The original engram will thus get sculpted and competing ones diminished or have their recall reduced in the current context until. Eventually, recall of the engram will not be subject to interference, and so it will not need the hippocampus and will not be subject to retrograde amnesia (Chapter 7).

On this view, SWS will involve replay and creation of an adjustable state for specific cell assemblies with a tendency to at least relative strengthening of wanted engrams. REM can then take over, with its theta peaks retaining the wanted engrams and "reactivations in the hippocampus [in] ... the troughs of the EEG theta rhythm help[ing] erase superfluous episodic memory information ... once a task becomes familiar".[R11, p. 88]

What is sleep for?

At the start of the chapter, I suggested that 'What is sleep for?' is a trap like 'What is an emotion?'. These questions assume that evolution produced only one result and intended to do so. It is better to think in terms of an evolutionary history driven by events and adaptations to them. From this perspective, we can see sleep, in the sense of torpor, evolving because it increased safety and allowed rest.

Given torpor as a starting point, sleep provides a distinct environment in which quite unrelated processes can party. It lacks the immediate constraints of the real world on the output of the processes. Simple rest enables the many processes in sleep that can be viewed as repairing the damage we have done to ourselves during the day. The release of growth hormone during early SWS could clearly help synaptic alterations – but will be affecting the whole body and not just the brain. Note that the evolution of sleep processes still depends ultimately on reproductive advantage expressed during waking. The key thing to remember is that sleep contains many processes that evolved, each driven by its own chance mutations and functional necessities.

SWS appears to have evolved at least in part to allow reconsolidation. The evolution of SWS appears to have provided a necessary foundation on which REM could build, with REM evolving at least in part to allow enhancement. Linked functions would explain why REM can only occur after a period of SWS. The processes we see in REM sleep also occur during wake (and appear essential for some memory encoding) – but their effect may be different when preceded by SWS and its reactivation of firing patterns. The weak effects of REM deprivation on declarative memories could be because theta occurs outside sleep (e.g., during encoding). But apparently weak effects will be observed if the process involved is enhancement in the sense of somewhat altering signal-noise ratios – and of converting malleable diffuse engrams to more rigid schemas. Such higher order changes will only be observable for particular memories and over the long term (at least ten years in the case of H.M.). Importantly, Walker's theory suggests that consolidation as such occurs in wake and so it is only subtler off-line processing resulting in reconsolidation and enhancement that requires sleep. Here the main function of SWS and REM taken together may be to reduce the load on memory storage and reduce the chances of catastrophic forgetting (to which pure Hebbian networks are prone).[R18] We will look in more detail at how information gets coded (as a percept or a memory) in the next chapter – where we will aim to put our detailed mental mechanics together into a picture of your mind machine.

Sleep, emotion, and mental disorder

My focus in this chapter has been on the possible complex role of SWS and REM in declarative memory. But, arguably, very similar processes are used, and better established as being more important, for procedural memory.[R31] I also glossed earlier over the importance of sleep processes to reduce neurodegeneration[R19] and improve mental health[R8] – where it plays a substantial role in the habituation of anxiety.

With the habituation of anxiety, it appears that under normal circumstances, REM results in habituation (potentially via LTD in the down phases of theta) – essentially enhancing your positive memories. However, with clinical anxiety, its strength appears to result in your negative memories being enhanced (likely by LTP in the up phases of theta). This accounts for the blocking of REM sleep by antidepressants and the brief antidepressant effects of sleep deprivation. The picture has added complexity, with some mental disorders involving disturbed sleep as both a cause and a symptom (for example Riemann et al).[R23] We will look at this further in Chapter 10 but one clear message is that, for the bulk of us, whether it is to improve mood, concentration, retain learned material for high exam grades, improve your competition diving routine, or support your immune system, you should aim to sleep well.

From my decades of experience as a course advisor to students, I recommend establishing good 'sleep hygiene' before you have any signs of problems, not once the damage has been done to your academic record (or job CV) or even tipped you over into an anxiety disorder. An internet search on 'sleep hygiene' should tell you what to do. But you need to remember that you will be trying to bring a range of processes into line – each of them as stubborn as the one that controls jet lag. Also, while some things (like a warm bed in a cool bedroom to help you relax and then go quickly into SWS) are easy to set up, others, like forcing yourself into a regular routine, can be really hard against social pressure.[R8] These are often best focused on one at a time. Above all, you will need patience (more so even than with jet lag). You are retraining ancient evolved routines and linked cycles (like SWS-REM). Not only will the new changes take time to bed in (when you can add the next such change), but they need to be in place for quite a while before you start feeling the full benefits.

Of course, if your disturbed sleep is the result of chronic pain or some other disorder,[R23] you should seek help for that. (Exam anxiety is a standard problem for many students). But, even then, sleep hygiene is likely to help.

Notes

1 Usually, parents have learned how to cope by the time they have second. One device my wife and I used with the arrival of our second child was sleeping at opposite ends of the house, with the baby alternating ends from day to day. This gave me a good sleep and my wife a reasonable sleep every second night despite the fact that she was breastfeeding.
2 Hz = cycles per second.
3 There are none marked in the original, but I think I see an α burst 2/3rds of the way along the trace.
4 Values reported earlier imply only moderate recovery of lost Stage 4 sleep and even less REM. But my calculations are for SWS as a whole (i.e., Stage 3 + Stage 4) and are not limited to Stage 4. Also, previous values are limited to the three recording days since the initial recording stopped at day 3. I have estimated day 4 values from the regression line and included them.
5 The behavioural description in the EEG report[R10] differs markedly from this.
6 I had to look this up! Astereognosis is the inability to identify objects by feeling only, without input from the visual system.
7 McNaughton is a single clan, so I must be related to Bruce if you go back far enough. But, when we have met, we have always been too busy discussing hippocampal cell firing to check this out. Among other fun experiments, he has recorded from rats' place cells on their fourth and ninth days in space on the shuttle *Columbia*.[R12]
8 For a highly detailed argument that "place cells" do not represent place and that the hippocampus controls interference for memories (spatial and otherwise) in other, particularly cortical, structures see McNaughton and Bannerman.[R14]

9 I have spent over 50 years researching theta rhythm using theories that have only recently become somewhat accepted. So, my views here should not be taken as generally accepted – but, of course, I think I have always been right.

10 Worryingly, this paralysis can suddenly occur at any time in a rare condition where you basically fall down while your dream appears as a hallucination superimposed on the traffic crossing or wherever you were at the time.

11 Strictly in terms of numerical definitions, absent some history, we have a problem. Human EEG theta is 4–6 Hz. This is often stretched to 4–7 Hz. Human EEG alpha is 8–12 Hz. But hippocampal rhythmicity in cats and rabbits is 4–6 Hz and so hippocampal rhythmicity was called "theta". Unfortunately, rat hippocampal rhythmicity runs 5–12 (or more) Hz but historically has been called theta. Case Vanderwolf suggested hippocampal Rhythmical Slow Activity should be called RSA. This was far too logical for rat traditionalists (who cares about human EEG?) so never caught on. Human hippocampus has a similar spread of theta/alpha frequencies to rat hippocampus but it looks like in two bands not one. Unfortunately, human cortical theta/alpha has multiple functional types. Even in rats, the lower frontal theta band is not linked to hippocampus while higher human-alpha may be under appropriate conditions.[R33]

12 If you care for your hearing, you do not want signal-amplifying headphones but noise-cancelling ones. This way you will need hearing aids a bit later in life.

13 People have trouble with double negatives. So, I have spelt out this idea in key chapters of a book[R15] and an even more detailed review [R14] Unless you are deeply interested in this area (or seduced by the consensus and want an antidote), ignore these, and just take my word for the idea that "amnesia" would be better called catastrophic hypermnesia (an excessive recall of incorrect possibilities that blocks retrieval of the correct item).[R18]

References

R1. Abraham, WC, How long will long-term potentiation last? *Philosophical Transactions of the Royal Society of London Series B-Biological Sciences*, 2003. **358**: 735–44.

R2. Allison, T & Cicchetti, DV, Sleep in mammals: Ecological and constitutional correlates. *Science*, 1976. **194**: 732–4.

R3. Barnes, DC & Wilson, DA, Slow-wave sleep-imposed replay modulates both strength and precision of memory. *The Journal of Neuroscience*, 2014. **34**: 5134–42.

R4. Bear, MF, How do memories leave their mark? *Nature*, 1997. **385**: 481–2.

R5. Beersma, DGM, Dijk, DJ, Blok, CGH, & Everhardus, I, REM sleep deprivation during 5 hours leads to an immediate REM sleep rebound and to suppression of non-REM sleep intensity. *Electroencephalography and Clinical Neurophysiology*, 1990. **76**: 114–22.

R6. Bollmann, L, Baracskay, P, Stella, F, & Csicsvari, J, Sleep stages antagonistically modulate reactivation drift. *Neuron*, 2025. **113**: 1446–59.e6.

R7. Ford, LDS, Morrison, AR, Mann, GL, Harris, JS, Yoo, L, & Ross, RJ, Sleep patterning and behaviour in cats with pontine lesions creating REM without atonia. *Journal of Sleep Research*, 1994. **3**: 233–40.

R8. Foster, RG, Peirson, SN, Wulff, K, Winnebeck, E, Vetter, C, & Roenneberg, T, Chapter Eleven – Sleep and circadian rhythm disruption in social jetlag and mental illness, in *Progress in molecular biology and translational science*, MU Gillette, Editor. 2013, Oxford: Academic Press. p. 325–46.

R9. Gahr, M, Schönfeldt-Lecuona, C, Spitzer, M, & Graf, H, Electroconvulsive therapy and posttraumatic stress disorder: First experience with conversation-based reactivation of traumatic memory contents and subsequent ECT-mediated impairment of reconsolidation. *The Journal of Neuropsychiatry and Clinical Neurosciences*, 2014. **26**: E38–9.
R10. Gulevich, G, Dement, W, & Johnson, L, Psychiatric and EEG observations on a case of prolonged (264 hours) wakefulness. *Archives of General Psychiatry*, 1966. **15**: 29–35.
R11. Inostroza, M & Born, J, Sleep for preserving and transforming episodic memory. *Annual Review of Neuroscience*, 2013. **36**: 79–102.
R12. Knierim, JJ, Poe, GR, & McNaughton, BL, Ensemble neural coding of place in zero-G, in *The Neurolab spacelab mission: Neuroscience research in space*, JCJ Buckey and JL Homick, Editors. 2003, Houston: NASA. p. 63–8.
R13. Lapierre, O & Montplaisir, J, Polysomnographic features of REM sleep behavior disorder: development of a scoring method. *Neurology*, 1992. **42**: 1371–4.
R14. McNaughton, N & Bannerman, D, The homogenous hippocampus: How hippocampal cells process available and potential goals. *Progress in Neurobiology*, 2024. **240**: 102653.
R15. McNaughton, N & Gray, JA, *The neuropsychology of anxiety: An enquiry into the functions of the septo-hippocampal system* (3rd ed.). 2024, Oxford: Oxford University Press. https://global.oup.com/academic/product/the-neuropsychology-of-anxiety-9780198843313?cc=es&lang=en&.
R16. McNaughton, N, Ruan, M, & Woodnorth, MA, Restoring theta-like rhythmicity in rats restores initial learning in the Morris water maze. *Hippocampus*, 2006. **16**: 1102–10.
R17. McNaughton, N & Vann, SD, Construction of complex memories via parallel distributed cortical-subcortical iterative integration. *Trends in Neurosciences*, 2022. **45**: 550–62.
R18. McNaughton, N & Wickens, J, Hebb, pandemonium and catastrophic hypermnesia: The hippocampus as a suppressor of inappropriate associations. *Cortex*, 2003. **39**: 1139–63.
R19. Parhizkar, S & Holtzman, DM, The night's watch: Exploring how sleep protects against neurodegeneration. *Neuron*, 2025. **113**: 817–37.
R20. Rauchs, G, Feyers, D, Landeau, B, Bastin, C, Luxen, A, Maquet, P, & Collette, F, Sleep contributes to the strengthening of some memories over others, depending on hippocampal activity at learning. *Journal of Neuroscience*, 2011. **31**: 2563–8.
R21. Raymond, CR, LTP forms 1, 2 and 3: Different mechanisms for the 'long' in long term potentiation. *Trends in Neurosciences*, 2007. **30**: 167–75.
R22. Rechtschaffen, A & Kales, A, *A manual of standardized terminology, techniques and scoring system for sleep stages of human subjects*, PH Service, Editor. 1968, Washington: US Government Printing Office.
R23. Riemann, D, Berger, M, & Voderholzer, U, Sleep and depression – Results from psychobiological studies: an overview. *Biological Psychology*, 2001. **57**: 67–103.
R24. Ross, JJ, Neurological findings after prolonged sleep deprivation. *Archives of Neurology*, 1965. **12**: 399–403.
R25. Sara, SJ, Reactivation, retrieval, replay and reconsolidation in and out of sleep: Connecting the dots. *Frontiers in Behavioral Neuroscience*, 2010. **4**: Article 185.

R26. Siegel, JM, The REM sleep-memory consolidation hypothesis. *Science*, 2001. **294**: 1058–63.
R27. Siegel, JM, Clues to the functions of mammalian sleep. *Nature*, 2005. **437**: 1264–71.
R28. Siegel, JM, Sleep viewed as a state of adaptive inactivity. *Nature Reviews Neurosciences*, 2009. **10**: 747–53.
R29. Skaggs, WE & McNaughton, BL, Replay of neuronal firing sequences in rat hippocampus during sleep following spatial experience. *Science*, 1996. **271**: 1870–3.
R30. Tononi, G & Cirelli, C, Perchance to prune. *Scientific American*, 2013. **309**: 34–9.
R31. Walker, MP, A refined model of sleep and the time course of memory formation. *Behavioral and Brain Sciences*, 2005. **28**: 51–104.
R32. Winson, J, The meaning of dreams. *Scientific American*, 1990. **263**: 42–8.
R33. Young, CK & McNaughton, N, Coupling of theta oscillations between anterior and posterior midline cortex and with the hippocampus in freely behaving rats. *Cerebral Cortex*, 2009. **19**: 24–40.

9 The big picture

Fitting the parts together

We now have a large parts list for the mind machine – from its building blocks to key processes: emotion, learning, memory, and sleep. When the machine is running all the parts are intertwined. It also engages complex mechanisms of perception and action that we have not covered yet. So how does the machine, as a whole, run? I will try and convince you that each part of each of these systems is basically simple. But the parts are contained within hierarchical layers that interact to generate the complex result that is your mind.

Our perceptual systems do not simply see the world 'as it is'. They also do not simply absorb what is 'out there'. They deduce and construct what they think should be there. This happens quickly, and seamlessly, so you do not realize that your picture of the world is an invented and distorted one. The systems distort the input they receive to improve the chances of detecting important objects at the expense of making mistakes with unimportant ones. On occasion, as a result, they detect objects that are not there – illusions. These distortions and inventions are "a feature not a bug". They promote your survival. They only seem odd when demonstrated in the laboratory. Indeed, as you saw in Experiment 2.3 different parts of you may be guided by different distortions of the same original input to achieve optimum results for their different purposes.

Early processing of visual information – contrast

Distortion and invention start with the earliest processing of visual information. Some interesting illusions result when we give the visual system simple inputs that it did not evolve to handle. Such illusions can be thought of as a psychologist's micro-electrode – illuminating the mechanical details of visual processing.

A classic illusion is the Hermann Grid (Figure 9.1). If you look at the repeating blocks you will see grey dots appear at the intersections of the lines except where you are directly looking. Move your eye a little and the dots

DOI: 10.4324/9781003465027-9

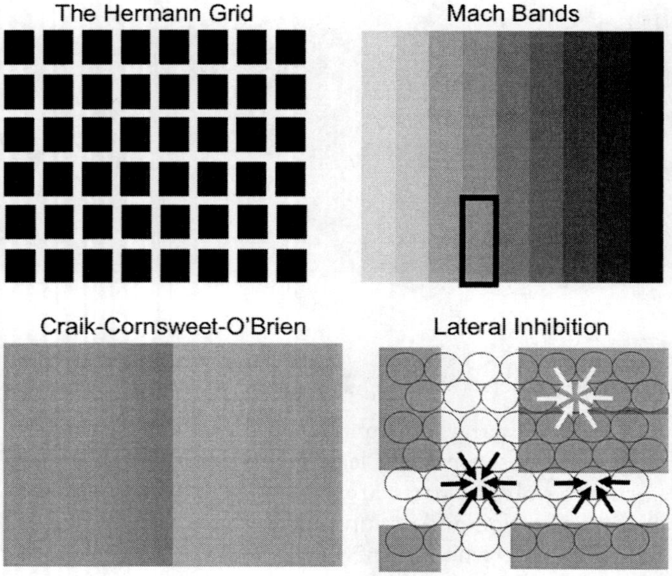

Figure 9.1 Contrast illusions. For details, see text.

seem to appear and disappear (with the centre of your vision still having no dots). With finer lines you would see dots in the centre.

Related to this are Mach Bands. If you look at the top 1/3rd of the vertical bands in Figure 9.1, the grey tone seems to be slightly scalloped within each band (looking slightly darker on the left than the right). But look at the bottom 1/3rd, where there is a box over the edges of one band and you can see that it is actually uniform grey. Use a pen or pencil to cover one of the bands and the two sides will now look more uniform. Like the Herman Grid, the Mach figure is changing perceived brightness close to edges. *As a result, the edges are clearer.*

One explanation for this kind of effect is that cells (in the retina or cortex)[1] have a 'centre-surround' organization that produces lateral inhibition (Figure 9.1). For example, with hexagonal organization, each cell would be surrounded by six other cells with inhibitory connections to it. If each inhibitory connection passed 1/6th of the positive effect of light on its cell (in real systems it's less than this), then the middle positive cell (at the intersection in the figure with six inhibitory arrows) would have no output. Its light input, taken to be 1 unit, would have subtracted from it (6 × 1/6) units of lateral inhibition. So, its output would be equivalent to zero input (as if it were in the dark). With the similarly wired cell, on the line to the right of the intersection, two of the inhibiting cells are in the dark (input 0), providing no inhibition, so the positive cell's output would be 1 – (4 × 1/6) = 1/3. That is, the output would appear lighter than for the cell at the intersection. That is what you see when you look – it appears darker at the intersection. Similar contrast calculations explain the Mach bands.

Note that the brain just gave itself a potential problem. Lateral inhibition will find edges very nicely. There will be greater output from a cell in the light whenever some of the inhibitory cells are in the dark. But our calculation (which is exaggerated and simplified) would 'see' both uniform white and uniform black as uniform black! That is, for white, $1-(6 \times 1/6) = 0$ (for all cells not adjacent to an edge); while, for black, $0-(6 \times 0) = 0$.

We can show that the brain fixes this contrast-enhancing distortion with some invention. Have a quick look at the Craik-Cornsweet-O'Brien rectangles in Figure 9.1. You should see a darkish grey rectangle on the left and a lighter one on the right. Now close one eye and hold a pen or pencil over the midline. Both rectangles will now be the same shade of grey. After detecting the edge between the rectangles (created by a narrow band where light grey shades to dark), your system has filled in away from the edge with the appropriate grey to correct for the loss of brightness information produced by lateral inhibition. In the real world, edges are important, and gradients of light and dark often result from uneven illumination not important features of objects and so can largely be ignored. Note that with cartoons like Calvin & Hobbes, you can be given pure black and white information and have no trouble seeing the picture.

Early processing of visual information – features

So, what does the system do with 'edges' once it has sharpened them? How does it get even to a cartoon level of representation? It starts off by extracting significant low-level features and then uses these to detect what may be parts of objects. Figure 9.2 shows David Marr's early computer-based model, which included neurally-realistic processing elements.[R14] Read the main body of the figure legend only if you are interested in the computational details.

Marr's computer program generated a high-level feature description of the input (including cartoon-like outlines) by clumping similar low-level features together. The model stopped before getting to object recognition as such (Marr suggested that going from outline elements to objects uses even higher order 'gestalt'[2] clumping and separating). I have superimposed on Marr's model a more recent framework proposed by Zhaoping Li.[R27] Three important changes from Marr's approach are: (1) separation of looking (adjusting your attentional focus) from seeing (identifying objects); (2) an emphasis on information compression; and (3) integration of bottom-up with top-down processing.

Looking is controlled by the primary visual cortex (V1) and peripheral retina. It can involve moving your attentional spotlight or moving your gaze direction. This shifts the centre of vision and can do so to an object that has not yet been identified. *Seeing* is controlled by higher visual areas and the

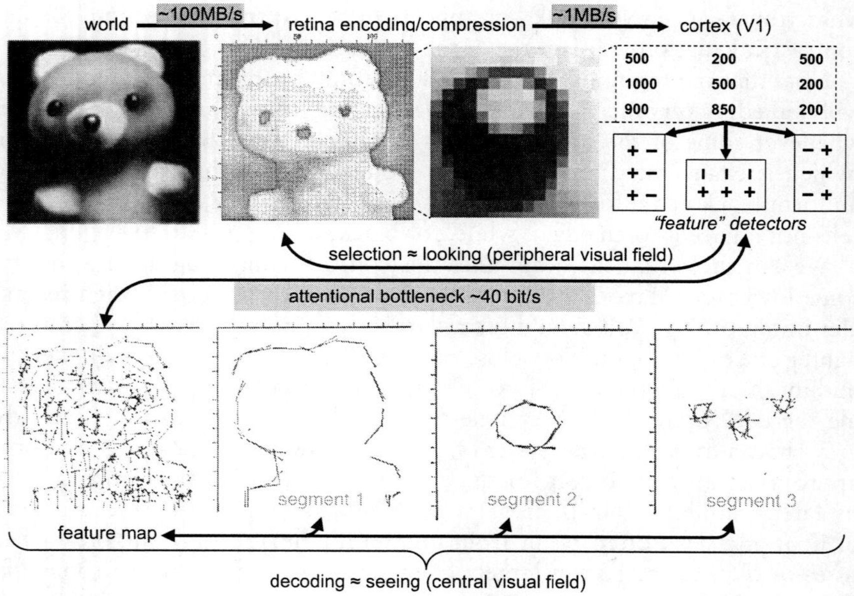

Figure 9.2 Early processing of visual information. The images are from David Marr's important initial demonstration of a computer-based vision system.[R14] His model was a series of steps where input to a 'retina' then led to a series of feature representations of ever higher abstraction. Superimposed on these is a more recent framework proposed by Zhaoping Li. This framework involves both much greater compression of information than Marr's and a major top-down input[R27] for a recent update see.[R28] Light enters the eye generating information at ~100 MB/s. This input is encoded and compressed and transferred to area V1 of the visual cortex at ~1 MB/s. Importantly, Zhaoping Li sees V1 as mainly selecting rather than processing information. Selection, in the sense of looking, often involves moving the point of view so that a key feature (e.g., the bear's nose) is in central vision (the fovea). This gaze direction is guided by the peripheral visual field and can orient to objects before they are recognized. The centre of the retina also selects information that is passed (through a ~40 bits/s attentional bottleneck) to higher areas for decoding (i.e., seeing). In contrast to Marr, Zhaoping Li sees major top-down information being returned from the higher areas to the central visual field to provide analysis-by-synthesis. Rather than the retina being functionally uniform, then, the periphery is for looking (without necessarily seeing) while the centre is for seeing. This system "selects a tiny fraction of visual input information for further processing and makes us almost blind to unattended inputs. We are intuitively unaware of this dramatic blindness since we do not know what it is like without this blindness. However, gating our perception, selection should take centre stage in vision".[R27, p. 1] (Images from Marr[R14] with permission of The Royal Society (UK); conveyed through Copyright Clearance Center, Inc.; improved bear photo from Frisby[R6] with permission of John P. Frisby).

central retina (fovea), which has much more tightly packed light-detecting cells than the periphery.

If you use digital images, you have almost certainly come across information compression. The JPEG file format is often used by digital cameras because it recodes information so the file does not have to store lots of individual values that are all the same. For example, the bear photo in Figure 9.2 is 240 KB as an original bitmap (one value for each pixel, matching how it is printed) but only 39 KB as a *high-quality* JPEG that can be decompressed to completely recover the original for printing or screen display. The JPEG compresses the large slabs of, for example, uniform black to a compact description. A low-quality JPEG compresses more and would be smaller still – but it would have lost some of the original information and when decompressed would result in a less clear image.

Your visual system does something akin to JPEG compression when it is looking for edges and ignoring uniform values (Figure 9.1). But it gets even greater compression by recoding the simple brightness image information into high-level object descriptions. These, like the edge calculations, can lose unimportant aspects of the original input. Critically the result is also much closer to recognition of the object. The visual system first filters the image with what are usually called 'feature detectors'. These are cells wired to react most strongly to a particular pattern of light and dark. The ones in Figure 9.2 would be called 'edge detectors' because *experimenters* see the pattern of + and – signs that way. But, in fact, they are not well tuned to detect edges, which would be done at higher levels of the system. They are most easily seen in primary visual cortex (V1), which is mostly involved in looking not seeing (but passes its information up to areas that do seeing, too).

Object identification involves the combination of bottom-up (what looks to be there) with top-down (what ought to be there) processing. Figure 9.3A shows a Kanizsa triangle. You should see a white triangle in the centre, with clear edges and a bright interior. Cover the bottom black blob and you will see that the edge of the triangle is an illusion. The brain has added the white triangle, top-down, to the bottom-up information coming from the eye. Figure 9.3B provides an image you can use in an experiment with your blind spot (the instructions for this are in the figure legend). There are two key messages from this experiment. First (as shown in Figure 9.3C) the nerves from your light-detecting cells are at the front of your retina and have to go through a small hole to get to the brain. There are no light-detecting cells there and so you are blind at that one spot. When you move the gap in the line to that spot, it disappears. (An octopus does not have this problem, Figure 9.3C.) Second, the missing formation is invented and filled in top-down. So, you see the line as being completed (looking like the line below). That is the simplest explanation your brain can provide, given the input it is receiving.

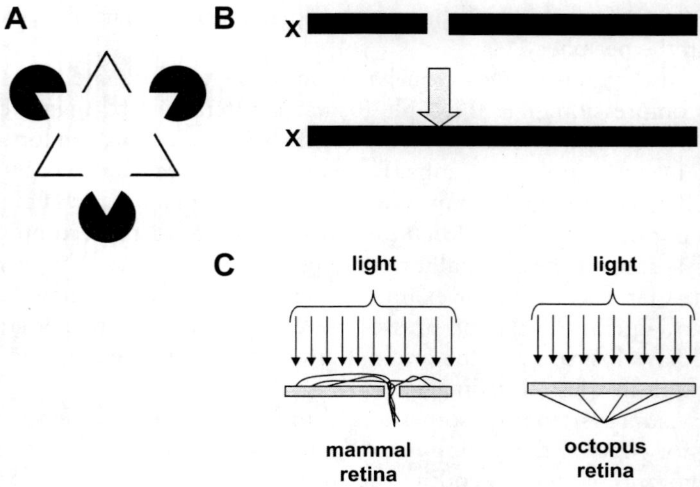

Figure 9.3 Invention – seeing what is not there. (A) The Kanizsa triangle. When you look at the image (particularly at some distance) you should see a white triangle in the centre. This triangle is usually seen as having clear edges and being brighter than its background. One possible cause is the top-down expectation of a triangle that fills in the object. However, it may also be a result of the centre-surround processing that produces, for example, Mach bands (Figure 9.1). It could be "that both types of explanation have a contribution to make"[R6, p. 139] particularly when contrast is high. (B) The blind spot. You may want to photocopy the figure for this – but try it with the book initially. Cover or close your left eye. Now, making sure you are looking directly at the top **x** with your right eye, bring the page closer to your face. At somewhere around 10 cm, you should see the white gap in the line disappear – and the line appear continuous. The disappearance shows that you have a blind spot in your eye that you don't normally notice. The filling in of the line shows that the brain can use top-down information to 'fill-in' during object recognition. (C) Mammal versus octopus. Your light-detecting cells have their wiring on the surface of the retina. This not only degrades the image a bit but also creates your blind spots where all the nerves dive through a small area in each of your retinas to get to your brain. Luckily, the spots are in a different part of the visual field for the two eyes. So, one eye can usually fill in for the other – or, as we just demonstrated, your top-down mechanisms can make a guess. Why is it wired in such a crazy way? That is just how it was when our eyes started evolving and despite the problems there was no way to go back. It definitely does not have to be like that if you start right. The octopus's eye has its wiring behind the retina, which makes much more sense to an engineer.

Later processing of visual information – what and where

After initial processing in area V1 and V2 objects are dealt with by two processing streams (Figure 9.4). One extracts what is there (and so helps you to work out what to do). The other extracts are where it is (and so where you need to direct your action). Keeping these aspects separate makes sense as

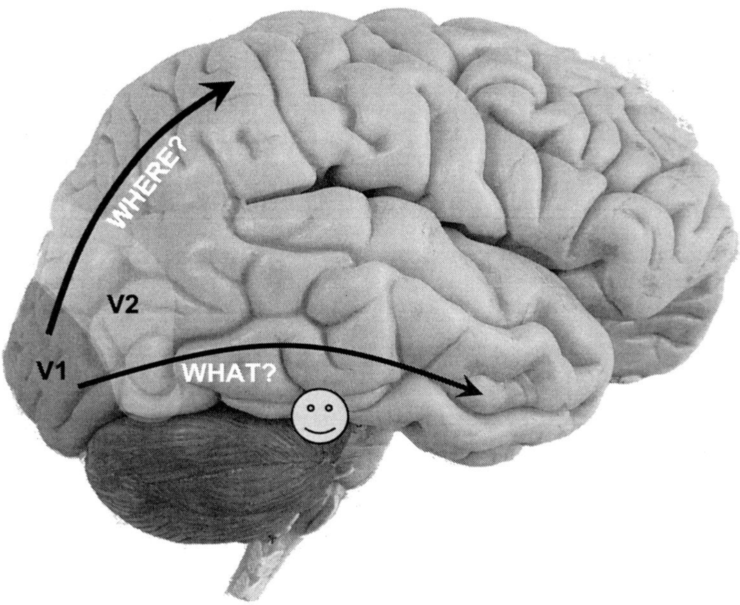

Figure 9.4 Extracting What and Where. Beyond area V1, information is processed in two streams: dorsal (Where) and ventral (What). The ventral one includes a dedicated face recognition area ☺ in the inferotemporal cortex.

they require quite different computations. So, this dorsal-ventral separation continues through to the control of actions and planning in the prefrontal cortex.

It likely does not feel reasonable that your knowledge of what and where should be separate. But you saw evidence of something like this in Experiment 2.3. The separation is most obvious in people who have damage to one or the other processing stream. They lose one or the other type of information. For an entertaining description of this kind of case, try reading Oliver Sacks's "The Man Who Mistook His Wife for a Hat, and Other Clinical Tales".[R22]

Later processing of visual information – specialization

There are also specialized processing areas within the streams. One smallish part of our ventral stream is involved in face recognition and an adjacent part is involved in visual recognition of words.[3] Interestingly, word recognition is generally lateralised to the language-dominant hemisphere (left side in right handers), while face recognition is opposite (right side in right handers).[R7]

I would now like you to do Experiment 1.1 [Part B]. (Do not go back and look at Part A again yet.) Look at the photos in Figure 9.5. Write down on a slip of paper which of Alastair and Ian you think Hamish looks more like.

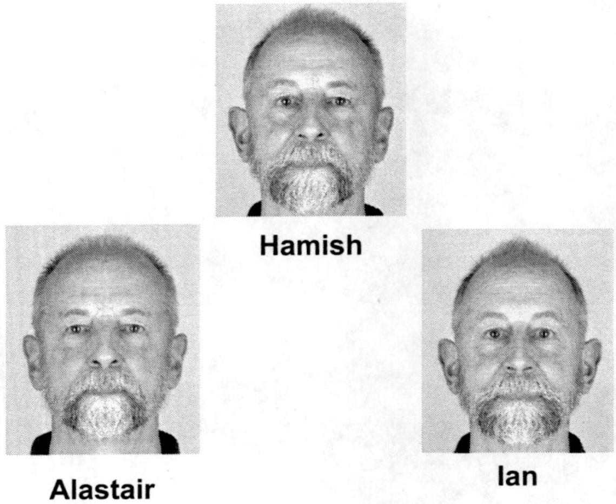

Figure 9.5 The fourth 'quadruplet photo'. Follow the instructions in the text.

If you are seeing the photos the way I do, you will think that Hamish looks more like Ian. Now go back and retrieve your slip of paper from Chapter 1. If you are seeing the photos the way I do, you will think that Neil looks more like Alastair. If you are left-handed, you may see the likenesses the other way round – or not see much difference.

Of course, the photos are not of quadruplets. They are all photos of me. 'Neil' is one of my standard passport photos. 'Hamish' is its mirror image. 'Alastair' is a copy of the left side of my face with its reflection added on the right. 'Ian' is a copy of the right side with its reflection added on the left. If you saw the photos the way I do, your choice of 'like' will be of the photo that matches the left side half-face of Neil or Hamish. Left-side visual information gets passed to the right side of the brain. That is where, as a right-hander, my face processing area should be dominant. If you want details of how to run this as a real experiment check this note.[4] There are more complex versions of this test where the 'chimeric' half-faces show different emotions.

With words and faces, a half-field of detailed information is clearly enough for recognition and so their lateralization is saving brain space (at a small cost in terms of not having an absolutely correct whole picture). However,

> a bias toward contralateral space remains evident at the highest levels of cortex and cognition [including vision, touch and motor control]. The left and right sides of vision appear to draw from separate cognitive resources and distinct attentional mechanisms. ... This organization also allows for focused spatial processing to be deployed largely in parallel between the left and right sides of space.
>
> (Brincat and Miller[R2, pp.1,5])

Nonetheless, there is signal mixing between the two particularly with motor control.

The key take-home message is that from the lowest to the highest levels, unexpected processes in quite separate systems deliver what we perceive as a unified and complete experience of the world. This perception is a mixture of distortion and invention. But, except in the laboratory or in ecologically rare circumstances, this 'false' picture will help you survive better than a strictly 'true' one would – and costs less to produce.

The perception-action cycle – spinal mechanisms

It's important to see your spinal cord as an extension of your brain. It can 'perceive' and 'take action' all by itself in the control of limbs. The action is, in a sense, goal directed, despite usually being termed 'reflexive'.

Some nerves from limbs in the spinal cord carry non-urgent sensory information (✪). (Mmm! Warm sun.) This goes directly to the brain (circuit 1, Figure 9.6) and not to the spinal neurons. But other nerves can carry urgent pain information (⚡) – as when you put your hand on a hot cooking element. If you then moved slowly, you could get seriously damaged. So, a withdrawal circuit (2, Figure 9.6) sends the pain information via an excitatory interneuron (○) to the relevant primary (α) motor neuron. Output from this neuron contracts the muscle, which moves your limb out of danger. In parallel, an inhibitory interneuron (●) blocks the *opposing* α motor neuron and so *relaxes* the muscle on the other side of the limb. Its contraction will not then slow withdrawal. The pain information is also sent, in parallel, to the brain. This allows for additional, slower, directed action (see below). I see this primary spinal circuit as having the goal of avoiding pain and damage.

The stretch circuit (3, Figure 9.6) is particularly interesting. No parallel signal is sent to the brain. The circuit pursues its own goal. It keeps your muscles at a set length despite occasional unexpected imposed loads.

The simplest stretch case is when you are trying to stand upright. If you sway forward, you could then fall flat on your face. It's important to correct the sway fast enough. Your muscle contains a spindle, which measures the change in length relative to a set value. Sway stretches the muscle and so the spindle, which sends a signal directly to the α neuron (and the brain). The α neuron contracts the muscle until the spindle is back at the length for which it is set. This is all very high speed. So, it beats gravity and often makes small enough quick adjustments that you won't notice unless you are carefully monitoring your sway. The correction process will continue (forward and/or back in small amounts using opposing muscles) as long as you try and remain vertical. One way to look at this is that *perception* of a small amount of sway leads to countervailing *action*. But both this perception and the action are unconscious. We will come to more complex examples of a perception-action cycle shortly.

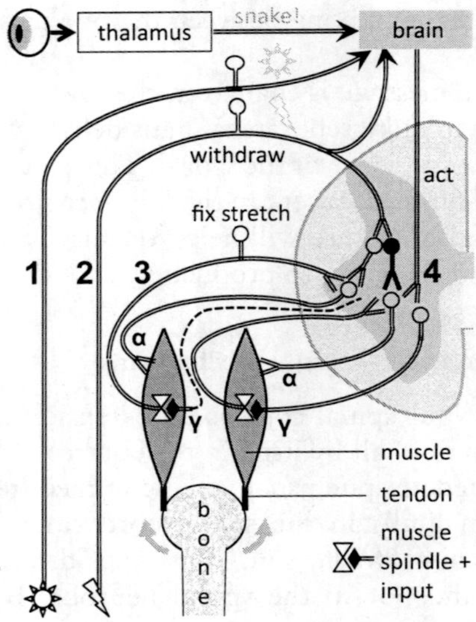

Figure 9.6 Peripheral motor control. See the text for explanations of the withdrawal reflex, stretch reflex, snake avoidance, and voluntary muscle contraction. **1, 2, 3, 4,** are sensory, withdrawal, stretch and voluntary action circuits, respectively. Inhibitory cell not shown on the right, for simplicity. Left side contraction produces leftward joint rotation (↻); Right side produces right (↺). ○ = excitatory neuron cell body; ● = inhibitory neuron cell body; α = alpha motor neuron (contracts muscle); γ = gamma motor neuron (contracts spindle). ☼ = sun on skin; ⚡ = pain.

Another example would be if you are holding your forearms level straight in front of you and someone puts a box onto them for you to hold up. Fast contraction to match the load means you will not drop the box. The length measured by the spindle means you only contract enough to hold your arms level. So, you hold the box up but do not throw it in the air. Central mechanisms then have time to make further adjustments if necessary.

The directed action circuit (4, Figure 9.6) is the last we need to talk about. An important part of this is the gamma motor neuron (γ). I ignored this when dealing with circuit 2 (showing it only as a dashed line) so as to keep the diagram simple. But it's equally important in both cases. The stretch reflex (circuit 3) does not need it. I have also omitted the inhibitory interneuron (●) in circuit 4 for the same reason. You can understand directed action without it; but it's equally important in both cases. I have also shown only one (directed action) input from the brain to the right-hand muscle. There is a parallel one to the left-hand muscle but this would operate separately (e.g., in parallel with withdrawal) – unless you are co-contracting opposing muscles (as for a body building display). So, focus on the bits shown to understand the main operation of the circuits; but include all the elements in a full model.

The basic motor side of directed action is simple. Output from your motor cortex activates both your α and γ motor neurons. The α contracts the target muscle (as it does with withdrawal and stretch). The γ motor neuron contracts the muscle spindle. This not only resets the desired length of the muscle but, if necessary, will indirectly engage the stretch reflex for as long as it takes to get the muscle to that length and keep it there. So, this maintains the effect of a brief input from the brain with a new goal length.

But, why would you contract that muscle? The example that I have included in the diagram is activation of the innate snake-detection circuit (Chapter 5). The simplest (and so fastest), thalamic, level of snake detection requires crude withdrawal in much the same way as pain would. So, all the brain has to do is activate the same α motor neurones as would produce an equivalent pain-withdrawal reflex.

Higher-level motor control is more complex. My emphasis, so far, here has been on the spinal cord doing things urgently, while keeping the central circuits informed. But spinal actions are necessarily simple. Fast, relatively undirected, withdrawal is an appropriate initial response to perception of a snake-like object. But, after this, more sophisticated avoidance strategies may be more adaptive (especially if the snake detection was a mistake) even if slower to execute (Experiment 9.1). As we move to the higher levels, perception and action become more entangled. When dealing with this, you may find it useful to return to these basics and think of the higher levels as operating in a very similar fashion – but with more depths of layering and with increased cognitive complexity simply because of the number of processing stages involved.

As we proceed bear in mind that complex perception does not entail complex action. Detection of a snake is quite complex but may require only simple withdrawal. Detection of a warning light in the cockpit of a passenger jet is simple. But it may then require a complex extended suite of actions with a high level of cognitive control. So, we will be moving up perceptual and motor hierarchies that are interconnected across many levels.

Experiment 9.1 Comparing the stretch reflex and directed action

To do this experiment, you will need the help of a couple of friends, a smartphone or digital camera that can record video, a chair without arms, and a small hammer. Having a tripod for the phone/camera could help too. For the analysis, you will need a computer (or possibly smartphone) with the VLC Media Player (free downloads are available for Windows, Android via Google Play, and in the Apple Store) or some other video player that can play back one frame at a time.

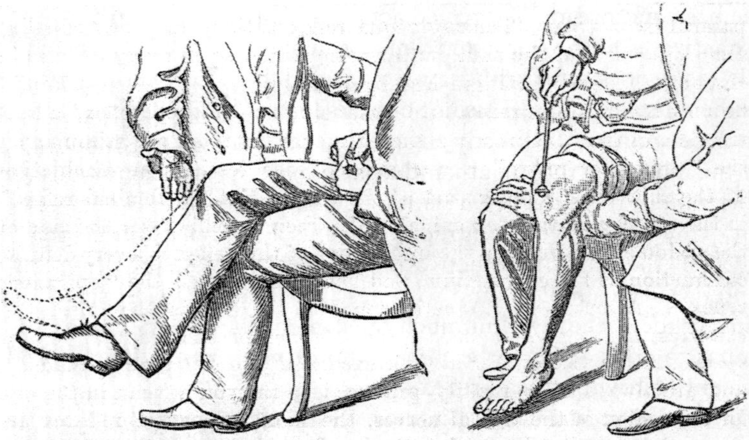

FIG. 1.—THE KNEE-JERK.
The dotted line indicates the movement which follows the blow on the patellar tendon.

FIG. 2.—THE KNEE-JERK.
Method of obtaining it when it is not readily produced in the ordinary way.

Setting up. Place the chair in a good-sized open space. Get one of your friends to volunteer to be hit on the knee by you. If you are right-handed, set up the other friend so they can hold (or activate if it is on a tripod) the phone/camera on the left of the chair. Set up your volunteer in the chair with their leg crossed as in Fig. 1 of Gowers 1886. Now see if you can elicit a 'knee jerk reflex' by hitting the patellar tendon. To find this, press with your finger on the kneecap (patella) and search for the large tendon that runs downward from it. It's important to hit the middle portion of the tendon. Too high or too low will give poor results. Try with the edge of your hand (as per Fig. 1). If that does not work, try using a small hammer (best with some cloth tied over the head to soften it). If the volunteer cannot cross their legs, or a reflex is not easily produced, try the method of Fig. 2. When you have a reliable result, make sure you can get a good recording of it from the angle shown in Fig. 1 that captures the starting movement of the leg. Check the settings on the video. Your frame speed should be at least FHD (60 fps, ~17 ms/frame), higher is better. For more accurate timing you could use slow motion (240 fps, ~4 ms/frame). This will need fewer samples to get clear results; but your files will be large (~1 MB/s of recording).

Running the experiment. Ask your volunteer to close their eyes. When recording has started, elicit the reflex. Stop recording, allow the volunteer to open their eyes. If the reflex or recording failed in some way, delete the file. Repeat until you have ten good recordings. Now tell your volunteer that you will hit them on top of their knee the same way you hit the tendon and that they are to voluntarily produce a knee jerk as fast as they can. Collect ten samples as for the reflex run. Then get a final ten samples of the reflex. You have an ABA design here, which counterbalances any effect of time/learning on the values obtained in the A phases.

Analysing the experiment. One at a time, load your files into VLC and run (▶) until it is just before when you hit the knee, then pause (II). Now step through frame by frame (press the 'E' key repeatedly) until the hammer hits the knee. Now, counting the number of E presses, step through to the point that you see the leg starting to move. Stop and record the number of presses (I recommend doing this in a spreadsheet like XL). Repeat this for all your 30 files. If using a spreadsheet, save the file, close, and reopen. Now multiply each value by 1,000 and divide by the frame rate you set for your video. This gives you the response time in milliseconds. Average each block of ten values. The two reflex averages should be very similar and are the time it has taken nerve impulses to pass all the way round the stretch circuit (3 in Figure 9.6). If they are not similar, you need either a higher frame rate or to collect more samples. Once they are similar, I predict that you will find that the top-of-the-knee response time is larger than theirs. The difference between it and them is the extra time taken for impulses to go to the cortex (circuit 1), activate the relevant instructed machinery, and send a signal back down to the α motor neuron (circuit 4).

The perception-action cycle – central mechanisms

Circuit 4 (Figure 9.6) carries the output of cells in the primary motor cortex (M, filled black dots, Figure 9.7). M is distinct in terms of cell type and often referred to as Brodmann Area (BA) 4. M is most easily thought of as being organised like a map of the body. The left-hand inset of Figure 9.7 illustrates the small numbers of the cells controlling unimportant body parts and large numbers controlling complex parts. Recent analysis has these effector-specific zones alternating with action-control zones (right-hand inset) Immediately beside M, on the other side of the central sulcus, is primary sensory cortex (S, horizontal lines, Figure 9.7, BA3). This has a very similar body mapping to M. M and S are reciprocally connected.

There is a hierarchy of motor control areas (see also Figure 4.7), with the higher levels evolutionarily older (last 2 sections of Chapter 4). All levels are reciprocally connected (⬅➡) to adjacent levels. Function gets more complex towards the front of the head. BA6 (pre-motor cortex, P) is parallel to BA4. It can be thought of as processing the next action that BA4 will execute. BA8 is parallel to BA6. It is usually referred to as the frontal eye fields because of its role in gaze control. However, I like to think of it as pre-premotor (PP). Eye movement to check a target will usually precede the planning (P) of

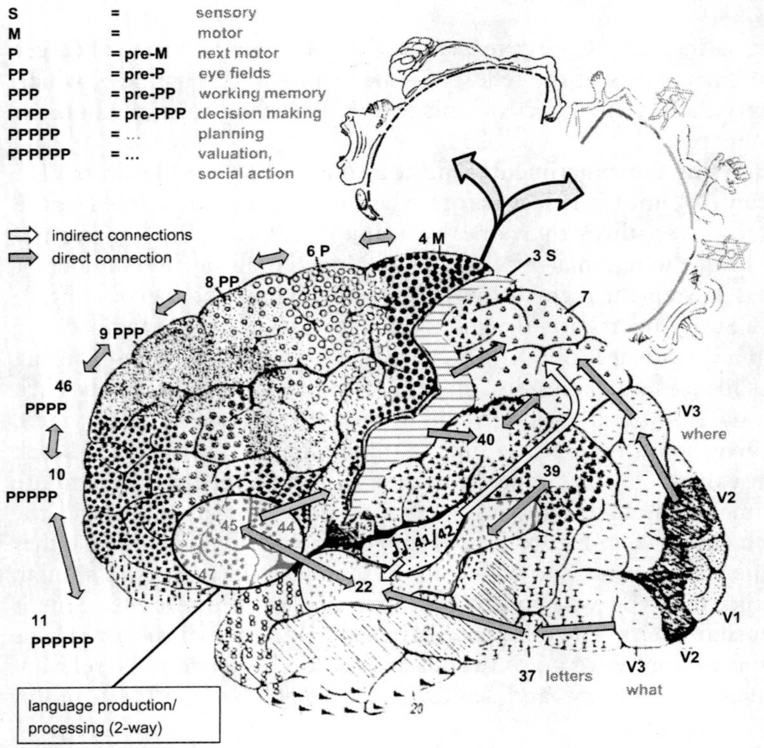

Figure 9.7 Perception and action – central control. Brodmann's cytoarchitectonic map of the lateral cortex [adapted from https://commons.wikimedia.org/wiki/File:Brodmann_Cytoarchitectonics_35.png public domain] with the hierarchical organization of action control superimposed. Brodmann's Area (BA) numbering allows a consistent identification of brain parts.[8] The primary motor cortex (M) is BA4. For other details, see text. **Inset:** a cross-sectional view of BA4, showing the relative amounts of it devoted to the different body parts and their organisation – left, classic homunculus proposed by Wilder Penfield[R20]; right, modern integrate-isolate model proposed by Evan Gordon et al.[R9] Images from Evan Gordon et al.[R9] licensed under a Creative Commons Attribution 4.0 International License, http://creativecommons.org/licenses/by/4.0/. Colours converted to patterns.

movement towards it. BA9 is parallel to BA8. It is usually seen as involved in working memory. But I like to think of it as pre-PP. It 'holds in mind' things like the position that the eyes will have to move to.[R8] BA46 is not clearly parallel with BA9. It is thought to be involved in decision-making – and, yes, I would see this as simply pre-PPP. BA10 is next in sequence, linked to planning/pre-decision making; i.e., pre-PPPP. BA11 engages in evaluation, emotional control, and elements of social action – all of which can be seen as necessary for planning (so pre-PPPPP).

Cortex above the principal sulcus is dorsal (where). Cortex below the principal sulcus is ventral (what) trend. This is a logical extension of their separation in posterior cortex. For our current purposes, they can be seen as working together – but with quite distinct results if they are damaged.

The perception-action cycle – language

You may have noticed that, as we went from BA4 to the front of the brain, the BA areas got shorter and thicker. (BA4, BA6, BA8, and to some extent BA9 are wedge-shaped, the rest are more rectangular.) As a result, there are areas below and behind BA46, BA10, and BA11 that we have not accounted for. A cluster of these in the dominant (usually left) hemisphere is involved in language production.

In Figure 9.7, I have circled what is called Broca's area (BA44, BA45) with an adjacent language-related area (BA47). This language production zone is connected (➡) to the part of BA4 that controls vocalization. Importantly, it is also connected to Wernicke's area (BA22) in the posterior cortex. Broca's and Wernicke's areas work together in language production *and* perception. Damage to either area can produce aphasia. Speech production involves planning of complex sequences, aligning it with the functions of BA10 and BA11. It also requires motor control of the vocal output (aligning it with BA4). Speech perception, in the basic case, involves auditory input (♫). Simple input to primary auditory cortex (BA41, BA42) undergoes detailed processing on its indirect (⇨) way to BA22. BA22 also receives visual input decoded as letters by BA37 (in the ventral/what visual stream). BA22 is also reciprocally connected with BA39 (the angular gyrus), which can be thought of as an extension of Wernicke's area and can be responsible for dyslexia and semantic aphasia.

I have focused on language as a special case of ventral 'what' object perception and its translation into action. There are of course a large number of related mechanisms even in the left hemisphere, with others on the right. For example, on the right-hand side of the brain BA39 seems to be involved in body image and, as we discussed earlier, BA37 is involved in face recognition. I will ignore other left-right hemisphere functional differences – and leave you to tackle their details later if you wish. Here, I want to focus on general principles that should help you do so. It's time to move on to a similarly simplified look at the dorsal stream and its detection of 'where' things are.

The perception-action cycle – location

While the dorsal trend of frontal cortex deals with the location of the goals that are identified by the ventral trend, it requires highly processed perceptual input. Determining spatial location is complex. In addition, such perception can only control action if it is linked to the current known positions of your body parts.

But from the point of view of mental mechanics, the organization of location processing is logical and fairly simple. The key target for location processing of incoming simple perceptual information is the superior parietal cortex, particularly BA7 (Figure 9.7). BA7 locates objects in space using sensory information and relates these to body part locations via proprioceptive information.

Body part information is already mapped by primary somatosensory cortex (S, BA3, Figure 9.7) and so this is transferred (➡) directly to BA7. The most important source of sensory spatial information is the visual system. This needs a progression of dorsal areas (V1➡V2➡V3➡) to extract location information that can then be used in, for example, reaching tasks. Auditory information from BA41/42 also undergoes detailed processing on its indirect (⇨) way to BA7.

The idea of reaching needs a bit of unpacking. BA40 is important for language processing, particularly reading. It is connected to a specific part of BA7, the lateral intraparietal cortex, which is involved in the control of eye movements (among other things). As you would expect BA40 works together with BA37 for this language function. You can think of eye movement during reading as reaching for words or letters.

There are two key take-home messages here. The first is that my simple overview of cortical processing has glossed over a number of cases where the details and processes are more complex. But when you dig into these you should find they follow similar mental mechanics to what I have discussed here. The second is that I have presented cases where adjacent areas are directly (or closely) interconnected. But you should note, for example, that despite both being linked to language, and side-by-side in the brain, BA39 and BA40 are not interconnected and are parts of quite distinct networks. Conversely, particularly to support working/active memory, frontal areas are reciprocally connected with posterior ones. I have not shown these connections (except for the language areas). But you should think of the Brodmann map as having, underneath it, a vast cat's cradle of interconnections.

The perception-action cycle – why

The lateral aspects of frontal cortex (including the parts that go over into mesial cortex) are, as we have seen, set up to determine what something is and where it is. But if you detect an object, should you withdraw from it or approach it? To determine the full nature of a goal, you need information about not only 'what' and 'where' but 'why'. For this you need not only the dorsal and ventral cortical trends but also what I call the mesial trend.

For our current purposes this is largely the cingulate cortex (BA25, BA24/33, BA23, BA29/30 in Figure 9.8) but also includes a range of other 'limbic' areas. Connections are shown from the anterior cingulate (BA24) to mesial parts of the dorsolateral frontal areas. But BA24 also connects to the same BA areas on the cortical surface shown in Figure 9.7. The orbital frontal cortex (OFC) and anterior cingulate (ACC) can be viewed as the cortical

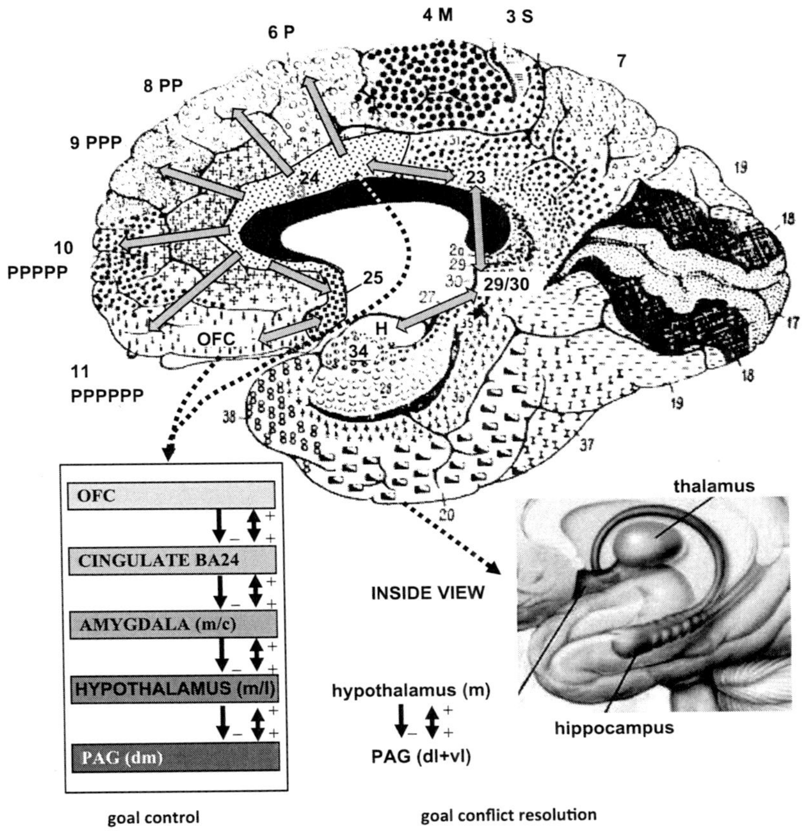

Figure 9.8 Emotion – central control. Brodmann's cytoarchitectonic map of the mesial cortex (down between the two hemispheres) [adapted from https://commons.wikimedia.org/wiki/File:Brodmann_Cytoarchitectonics_24.png public domain]. Action areas extend over from the outside of the hemisphere and are labelled as in Figure 9.7. Their connections with cingulate cortex (BA25, BA24/33, BA23, BA29/30) are shown. OFC = orbital frontal cortex; H = hippocampus (mostly obscured by temporal cortex including the entorhinal area (BA34); PAG = periaqueductal grey. **Inset:** OFC and cingulate subcortical connections for goal control. The primary motor cortex (M) is BA4. For other details, see text. INSIDE VIEW: as for Brodmann areas but with cortex shown as transparent (https://www.wpclipart.com/medical/anatomy/brain/human_brain_drawing.jpg.html public domain, see also Figure 4.6) and with connections from hippo campus to hypothalamus and PAG shown.

extensions of the subcortical areas that control attraction to and repulsion from[5] goals.

The lowest (fastest) level of subcortical goal control is the periaqueductal grey (PAG).[R25] With repulsion, for example, the PAG reacts to signals of suffocation by producing panic. The hypothalamus, at the next level, deals with more complex, but immediate, reactions (e.g., fight/flight). The amygdala deals with less urgent reactions (e.g., avoidance[R13]) triggered initially by fast thalamic input and then adjusted by slower (more sophisticated) posterior cortex input. BA24 and then OFC control ever more anticipatory behaviour.

An important feature of this control is that it is emotionally loaded. The anterior mid-cingulate appears to integrate negative affect, pain, and cognition linking information about reinforcers to control of goal-directed behaviour.[R24] "An increasing number of studies have demonstrated the involvement of the ACC in the emotional component of pain and its comorbidity with emotional disorders such as anxiety and depression".[R12, p. 1] That is, it appears to register (generate feelings of) and integrate both physical and emotional pain. "Cingulotomy mostly occludes pain-related emotions, but does not affect the patient's ability to identify the intensity and localization of the noxious stimulus".[R26, p. 200]

The integration of affect, pain, and cognition is particularly pronounced in the role of ACC in production of obsessions and their accompanying compulsions. We will discuss obsessive-compulsive disorder (OCD) with other mental disorders in the next chapter. But, in the context of higher order anticipation of negative events, obsessive behaviour (like hand washing, wearing a mask during a COVID epidemic, checking the door is locked, etc.) in the absence of a safety signal is adaptive. You cannot tell if viruses or bacteria are there. But if you regularly (but not excessively) perform the appropriate guarding actions you will increase your chance of survival and (more important from a Darwinian point of view) the production and survival of your children. In relation to OCD, Chapter 10 details how the cingulate appears to generate obsessions that, in interaction with the caudate nucleus (part of the basal ganglia involved in motor control), produce adaptive (and maladaptive) compulsive behaviour.

From parts to whole

We have now looked, separately, at key superficially distinct operations of the mind machine (emotion, learning, memory, sleep); and, in this chapter, different levels and trends of processing that support perception-action cycles. The levels allow an adaptive choice between (horizontally) parallel quick and dirty versus slow and sophisticated systems. The what, where, and why streams are similarly (but vertically) parallel.

Earlier, we contrasted conscious vision with unconscious blindsight. These involved distinct constructions and distortions of perceptual input to optimize performance in the two cases. We also contrasted unconscious control of leg movement (via the stretch reflex) with conscious (voluntary) control.

In contrast to my piecemeal approach to brain operation and emphasis on conscious/unconscious differences, I am sure that you feel that your mind operates as an integrated whole; not an independent set of parts. It's also clear that the key element of action control, a goal, must *combine* what, where, and why – without any one of these it cannot be a goal. So, now we need to look at how the different parts of the machine mesh during real life (as opposed to lab experiment) operation.

At the simple system level some integration is achieved by the relevant levels and trends being interconnected. This allows, for example, activation to spread between subcortex and cortex – with inhibitory connections determining which level generates behavioural output (Figure 9.8, goal control inset).

But at the complex system level integration is achieved via large-scale circuits. These can operate primarily between the cortex and subcortex or primarily between different parts of the cortex. We will look first at the Papez circuit as an example of the former; and then at the salience, fronto-parietal, and default mode networks (DMNs) as examples of the latter – and of interactions between the networks.

The Papez circuits

In 1937 James Papez described "a proposed mechanism of emotion".[R19] His description of this circuit started in the hypothalamus with the mamillary body (MB). This detects 'emotional colouring' that can be added to simple sensations transmitted by the thalamus to the cortex. The MB sends this 'stream of feeling' via the anterior thalamic nuclei (AMT, AVT, ADT) to the medial wall of the cerebral hemisphere, particularly the cingulate cortex (Figure 9.9, left side circuits). But, according to Papez, emotion can also originate in the cortex and pass (right-side circuits) to the hippocampal formation and then the MB. In both cases, activity can continue to circulate around the unidirectional loops connecting the structures. When looking at Figure 9.9, do not try and understand what the circuits are doing. As Papez suggested they play a role in emotion; but they are also involved in memory consolidation.[R17]

One key take-home message should be that what appears to be a complex set of loops is actually quite simple when you realize that there are small loops within the large ones and that the large loops are stacked in parallel (essentially reflecting the quick and dirty versus slow and sophisticated processing that we met earlier).

A second take-home message is that such looping circuits can produce immensely powerful, iterative, processing. Rather than needing a lengthy hierarchy of structures, deep processing can be achieved by continuously re-circulating and adjusting an original input until a much higher-level output form is achieved. This allows the formation of complex memories using simple Hebbian processes[R17] and the resolution of complex goal conflicts.[R15]

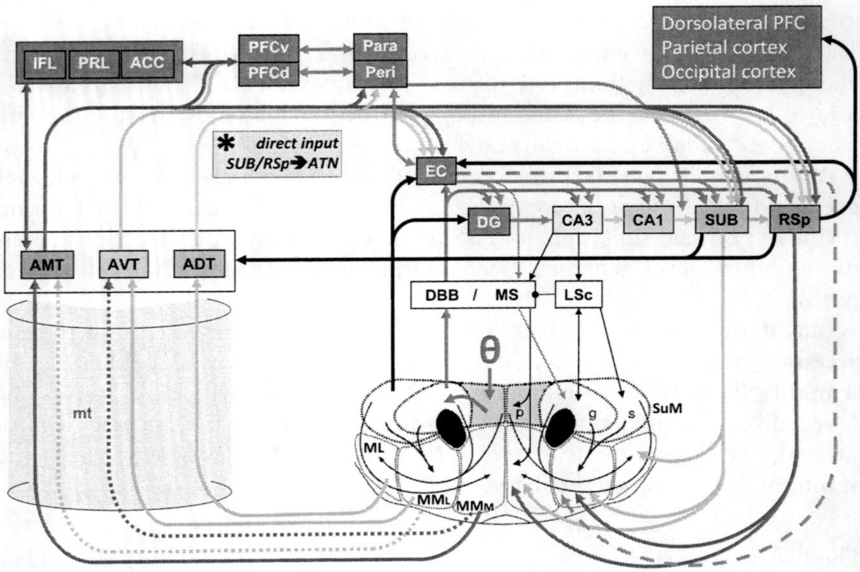

Figure 9.9 The Papez circuits. Overview of long and short loop connections from the hippocampus via mammillary area to the frontal cortex and back present in mammals ranging from rodents to primates [27, 50, 70, 121–126]. The mammillary bodies (ML, MM_L, MM_M) target prefrontal and anterior cingulate cortex, which target the hippocampal formation, completing the Papez circuit in onion-like, nested loops. The parts of the hippocampal formation (EC, DG, CA3, CA1, SUB and RSp) connect unidirectionally. Successive loops are similar, but outside loops have greater delays and more highly processed information. Abbreviations: ADT, AMT, AVT = anterior thalamus, dorsal, medial, ventral, respectively; DB/MS = medial septum; LS = lateral septum; IFL, PRL, ACC = medial limbic cortex; ML, MML, MMM = mamillary bodies, lateral, medial pars lateralis, medial pars medialis, respectively; mt = mammillothalamic tract; Para = parahippocampal cortex; Peri = perirhinal cortex; PFC = prefrontal cortex. Legend and figure (converted to grey-scale) simplified from Xiao and Zhang[R17] via CC BY license (http://creativecommons.org/licenses/by/4.0/).

The salience, fronto-parietal, and default mode networks

More global complex integration is achieved by the fact that the structures we have talked about in earlier sections are parts of at least five large-scale cognitive control networks: salience, cingulo-opercular, ventral attention, dorsal attention, and fronto-parietal. These are highly interconnected, multi-directional, networks and can be contrasted with the more unidirectional closed-loop Papez circuits (Figure 9.9). I will briefly describe two of them (salience and fronto-parietal) and then their relation to the DMN – which is active when they are not. This should give you a picture of their complexity. But you will not get a good idea of their overall functions since these are not yet known. All these networks involve a wide range of cortical structures;

and which components are active varies from task to task. In addition, the networks interact with each other.

The salience network is

> a suite of brain regions whose cortical hubs are the anterior cingulate and ventral anterior insular (i.e., fronto-insular) cortices. This network, which also includes nodes in the amygdala, hypothalamus, ventral striatum, thalamus, and specific brainstem nuclei, coactivates in response to diverse experimental tasks and conditions, suggesting a domain-general function.
>
> (Seeley[R23, p.9878])

The two hubs can be thought of as receiving visceral and autonomic feedback and then

> generating relevant visceral, autonomic, behavioral, and cognitive responses. Through interactions with each other, these regions could form a sort of information processing loop for representing and responding to homeostatically relevant internal *or external* stimuli and imbuing these stimuli with emotional weight.
>
> (Seeley[R23, p.9880, emphasis added])

They also contain special 'von Economo' neurons that are thought to support social functioning. (The physically close cingulo-opercular network was originally thought to be included in the salience network but does not have these special neurons and appears more concerned with continuing vigilance than sudden salience.)

The front-parietal network consists primarily of lateral prefrontal cortex and dorsal parietal cortex. That said, it includes multiple 'hubs' that flexibly alter their connectivities in relation to task demands. It occupies a central position interconnecting the other networks (cognitive, default, as well as the simpler sensory and motor ones).[R5] In a sense you can see it as coordinating activity among them.

The DMN is different. Its discovery

> revolutionized our understanding of the workings of the human brain. ... The DMN is a collection of distributed and interconnected brain regions that are typically suppressed when an individual is focused on external stimuli; however, in the absence of attention to external stimuli, the DMN switches or 'defaults' to internally focused thought processes, such as self-reflection, daydreaming, mind wandering, recall of personal experiences, and envisioning the future. The DMN is also hypothesized to be active during periods of 'rest' and quiet wakefulness, allowing us to reflect on our experiences, feelings, and social interactions.
>
> (Menon[R18, p. 2469])

It includes: dorsal and ventral parts of medial frontal cortex; posterior cingulate areas; anterior, medial and lateral temporal areas; the angular gyrus (see language section above); and, subcortically parts of the thalamus, medial septum, and nucleus accumbens.[R18]

The triple-network model[R18] combines these three networks (Figure 9.10). External events that are relevant to homeostasis generate sensory input that activates the salience network. This increases attention to external stimuli and sends input to the front-polar network that can result in network switching. Both the salience and front-polar networks send inhibitory inputs to suppress activity in the DMN. This blocks any ongoing internal narratives.

Once sensory input and the responses to it cease, the DMN is released from inhibition. This causes a rebound increase in activity that switches to an internal focus with resumption (or new generation) of internal narratives. Flexible switching between the background frames of these produces global state changes. These, in turn, can generate mind wandering.

Importantly, the lack of externally elicited brain activity is not replaced by a simple null state (or sleep). Instead, it is replaced by internally-generated activity. This supports "key cognitive functions subserved by the DMN, including self-reference, social cognition, episodic memory, and its role in personally relevant aspects of remembered experiences, language comprehension and semantic memory, and mind wandering".[R18, p. 2475]

The hard problem of consciousness[6]

I think mental mechanics accounts, in principle, for how your thoughts, feelings, and emotions work. The previous two sections suggest how all the parts

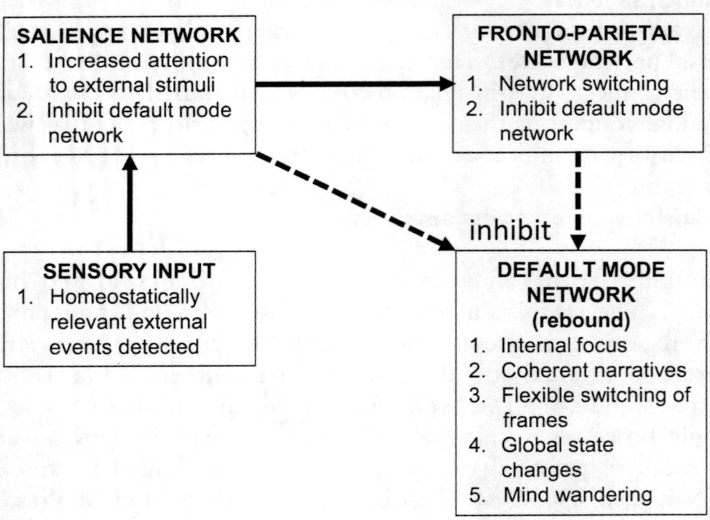

Figure 9.10 The triple-network model.[R18] For details see the text.

can come together to produce integrated action. This integration should include both conscious (in the sense of reportable) and unconscious (like blindsight[7]) processes.

But I have not tried to explain your 'consciousness' in the sense of your personal experience of being *you*. I have always doubted supposedly scientific theories of personal internal experience. I have no problem with distinguishing conscious/reportable from unconscious/unreportable processes; nor with the demonstrations (using brain imaging, particularly EEG) that some people (with complete locked-in syndrome) who appear in a vegetative state are actually aware of the world. They respond to instructions by imagining actions that they cannot take in reality. Cunning brain recordings then allow us to determine which action they are imagining. Indeed, development of brain-computer interfaces appears capable of giving these sufferers back some control of their physical world.[R21] But this only shows that they can hear and understand and, at least internally, respond. It gives us no idea of their actual experience. My wife and I agree on what is turquoise but not whether that is blue or green. So, if we agree about a particular object being green are we having the same experience – perhaps her red and her green (on the inside) are swapped compared to mine?

Given this, how can *you* tell if even *I* (as I write this) am conscious in the strong experiential sense as opposed to reportability via words or brain imaging or keyboard presses? This question is usually referred to as the hard problem of consciousness[R3] for an update, see Chalmers[R4]. See Arnold et al.,[R1] Graziano,[R10] and He,[R11] for criticism of a recent theory of consciousness as unscientific and add[R10,11] for a picture of general confusion.

Similar problems arise with 'free-will'. You would need a book to disentangle all the things this could mean and imply. But it is only a problem with syllogistic philosophy of the form:

- My brain activity is me.
- So, I am *just* a machine.
- So, you cannot blame me for anything.

But blame is not a concept in neuroscience. It's a social one. They are at different logical levels. That said, you can use mental mechanics to show that an action was the result of a brain seizure rather than conventional intent. You cannot use it to decide whether ordinary healthy actions should be subject to praise and blame.

To get a bit silly about this, it has been suggested that consciousness is a quantum field. This could be generated in microtubules or molecular effects at synapses. The nice thing about this hypothesis is that its consciousness is necessarily an overall (potentially even outside) the brain thing. On this view, your free will affects the otherwise random outputs of the quantum field, operating at a deterministic point in the decision pathway. Quantum theory says you cannot know which way the particle will jump. But perhaps free will (outside the universe defined by science) can make the decision. But, how is this *your* free-will (rather than just randomness in decision making). Well, obviously, your soul (which lives outside the physical universe) controls the 'random' processes. This would, of course, allow it to transmigrate. If you want to believe all of this, fine. But don't expect scientific backup as we cannot (by definition) measure discrete quantum effects on individual atoms, far less constellations of 10 billion neurons.

I think that mental mechanics allows you to understand how your mind works and what your mind is and how other people operate. But when it comes to moral, philosophical, and related issues you need to turn to the humanities, not science, that is why universities have them. Of course, as soon as someone comes up with a scientifically water-tight way of determining the nature of your conscious internal experience, mental mechanics can jump in and start explaining where it comes from.

This leaves one area to consider where mental mechanics has made major advances recently. The brain is fundamentally evolved and adaptive. But what happens to your mind when its mental mechanics go wrong or its environment does not match its adaptations? That is the focus of the next chapter.

Notes

1 There is lateral inhibition in the retina of Horseshoe Crabs (*Limulidae*), but the Hermann Grid effect is likely cortically generated in humans.
2 Gestalt is a German word that means something like pattern/whole/complete. In psychology, a gestalt is a whole percept that is more than the sum of the parts from which it is derived.
3 Both are in the fusiform gyrus, located on the bottom surface of the inferotemporal cortex.
4 Different people will see these images differently. Also, different faces have more or less asymmetry. To make images for an experiment on this get a series of passport-style photos as JPEG images (or cutout similar full-face images from pictures on the internet. Load them into a simple image program that allows a bit of editing (Paint will be on all Windows machines) and cut, rotate, and save the bits you need. They can then be put into a program like PowerPoint and organized to give the equivalent of my photos. It's likely simplest to label the photos something like A1, B1, C1; A2, B2, C2 and ask people, in each case "Is A more like B or more like C". Then count the number of times their choice is of the left half field (and make sure to note if they are left- or right-handed). It's best to only have one of each A image in a series (which is why I kept your exposures to mine separated by several chapters), but you can have two sets if you flip the A images to make a second set.
5 "Goal" often implies attraction. But to understand motivational systems, the concept of a negative goal (e.g. a predator) is useful in that it can be viewed as producing reactions just like those of a positive goal, except opposite in direction, see Figure 2 in McNaughton et al.[R16]
6 See also https://en.wikipedia.org/wiki/Hard_problem_of_consciousness.
7 Chapter 2 Perception.
8 BA numbers are good for an internet search – but put this in as "Brodmann Area …" or you may get a British Airways flight.

References

R1. Arnold, DH, Baxter, MG, Bekinschtein, TA, Bengio, Y, Bisley, JW, Browning, J, Buonomano, D, Carmel, D, Carrasco, M, Carruthers, P, Carter, O, Chang, DHF, Charest, I, Cherkaoui, M, Cleeremans, A, Cohen, MA, Corlett, PR, Christoff, K, Cumming, S, Cushing, CA, de Gelder, B, De Brigard, F, Dennett, DC, Dijkstra, N, Doerig, A, Dux, PE, Fleming, SM, Frankish, K, Frith, CD, Garfinkel, S, Goodale, MA, Gottlieb, J, Hanson, JR, Hassin, RR, Herzog, MH, Heyes, C, Hsieh, P-J, Hung, S-M, Kentridge, R, Knapen, T, Konstantinou, N, Kording, K, Kvamme, TL, Kwok, SC, Lanfranco, RC, Lau, H, LeDoux, J, Lee, ALF, Libedinsky, C, Lieberman, MD, Lin, Y-T, Liu, K-Y, Machizawa, MG, Martinez-Trujillo, J, Metcalfe, J, Michel, M, Miller, KD, Mitra, PP, Mobbs, D, Mok, RM, Morales, J, Mylopoulos, M, Odegaard, B, Or, CCF, Owen, AM, Pereplyotchik, D, Pestilli, F, Peters, MAK, Phillips, I, Rademaker, RL, Rahnev, D, Rees, G, Ringach, DL, Roskies, A, Schiller, D, Schurger, A, Schwarzkopf, DS, Scott, RB, Seitz, AR, Shepherd, J, Silvanto, J, Slagter, HA, Smith, BC, Solovey, G, Soto, D, Spiers, H, Stein, T, Taschereau-Dumouchel, V, Tong, F, Tse, PU, Vibell, J, Watzl, S, Webb, T, Weisberg, J, Wheatley, T, Wierzchoń, M, Wokke, ME, Yan, K, Klincewicz, M, Cheng, T, Schmitz, M, Sebastián, MÁ, Snyder, JS, & Concerned, IIT, *What makes a theory of consciousness unscientific?* Nature Neuroscience, 2025. **28**: 689-693

R2. Brincat, SL & Miller, EK, Cognitive independence and interactions between cerebral hemispheres. *Neuropsychologia*, 2025. **212**: 109153.

R3. Chalmers, DJ, Facing up to the problem of consciousness. *Journal of Consciousness Studies*, **2**: 200–19.

R4. Chalmers, DJ, How can we construct a science of consciousness? *Annals of the New York Academy of Sciences*, 2013. **1303**: 25–35.

R5. Cole, MW, Reynolds, JR, Power, JD, Repovs, G, Anticevic, A, & Braver, TS, Multi-task connectivity reveals flexible hubs for adaptive task control. *Nature Neuroscience*, 2013. **16**: 1348–55.

R6. Frisby, JP, *Seeing*. 1979, Oxford: Oxford University Press.

R7. Gerrits, R, Van der Haegen, L, Brysbaert, M, & Vingerhoets, G, Laterality for recognizing written words and faces in the fusiform gyrus covaries with language dominance. *Cortex*, 2019. **117**: 196–204.

R8. Goldman-Rakic, PS, Cellular basis of working memory. *Neuron*, 1995. **14**: 477–85.

R9. Gordon, EM, Chauvin, RJ, Van, AN, Rajesh, A, Nielsen, A, Newbold, DJ, Lynch, CJ, Seider, NA, Krimmel, SR, Scheidter, KM, Monk, J, Miller, RL, Metoki, A, Montez, DF, Zheng, A, Elbau, I, Madison, T, Nishino, T, Myers, MJ, Kaplan, S, Badke D'Andrea, C, Demeter, DV, Feigelis, M, Ramirez, JSB, Xu, T, Barch, DM, Smyser, CD, Rogers, CE, Zimmermann, J, Botteron, KN, Pruett, JR, Willie, JT, Brunner, P, Shimony, JS, Kay, BP, Marek, S, Norris, SA, Gratton, C, Sylvester, CM, Power, JD, Liston, C, Greene, DJ, Roland, JL, Petersen, SE, Raichle, ME, Laumann, TO, Fair, DA, & Dosenbach, NUF, A somato-cognitive action network alternates with effector regions in motor cortex. *Nature*, 2023. **617**: 351–9.

R10. Graziano, MSA, Illusionism big and small: Some options for explaining consciousness. *eNeuro*, 2024. **11**. Oct 29;11(10):ENEURO.0210-24.2024.

R11. He, BJ, Next frontiers in consciousness research. *Neuron*, 2023. **111**: 3150–3.

R12. Journee, SH, Mathis, VP, Fillinger, C, Veinante, P, & Yalcin, I, Janus effect of the anterior cingulate cortex: Pain and emotion. *Neuroscience & Biobehavioral Reviews*, 2023. **153**: 105362.
R13. LeDoux, JE, Emotion, memory and the brain. *Scientific American*, 1994. **270, June**: 50–9.
R14. Marr, D, Early processing of visual information. *Philosophical Transactions of the Royal Society of London. B, Biological Scien*ces, 1976. **275**: 483–519.
R15. McNaughton, N & Bannerman, D, The homogenous hippocampus: How hippocampal cells process available and potential goals. *Progress in Neurobiology*, 2024. 240, 102653.
R16. McNaughton, N, DeYoung, CG, & Corr, PJ, Approach/avoidance, in *Neuroimaging personality, social cognition and character*, JR Absher and J Cloutier, Editors. 2016, San Diego: Elsevier. p. 25–49.
R17. McNaughton, N & Vann, SD, Construction of complex memories via parallel distributed cortical-subcortical iterative integration. *Trends in Neurosciences*, 2022. **45**: 550–62.
R18. Menon, V, 20 Years of the default mode network: A review and synthesis. *Neuron*, 2023. **111**: 2469–87.
R19. Papez, JW, A proposed mechanism of emotion. *Archives of Neurological Psychiatry*, 1937. **38**: 725–43.
R20. Penfield, W & Rasmussen, T, *The cerebral cortex of man: A clinical study of localization of function*. 1957, New York: MacMillan.
R21. Rezvani, S, Hosseini-Zahraei, SH, Tootchi, A, Guger, C, Chaibakhsh, Y, Saberi, A, & Chaibakhsh, A, A review on the performance of brain-computer interface systems used for patients with locked-in and completely locked-in syndrome. *Cognitive Neurodynamics*, 2024. **18**: 1419–43.
R22. Sacks, O, *The man who mistook his wife for a hat, and other clinical tales*. 1985, London: Picador.
R23. Seeley, WW, The salience network: A neural system for perceiving and responding to homeostatic demands. *The Journal of Neuroscience*, 2019. **39**: 9878–82.
R24. Shackman, AJ, Salomons, TV, Slagter, HA, Fox, AS, Winter, JJ, & Davidson, RJ, The integration of negative affect, pain and cognitive control in the cingulate cortex. *Nature Reviews Neurosciences*, 2011. **12**: 154–67.
R25. Silva, C & McNaughton, N, Are periaqueductal gray and dorsal raphe the foundation of appetitive and aversive control? A comprehensive review. *Progress in Neurobiology*, 2019. **177**: 33–72.
R26. Xiao, X & Zhang, Y-Q, A new perspective on the anterior cingulate cortex and affective pain. *Neuroscience & Biobehavioral Reviews*, 2018. **90**: 200–11.
R27. Zhaoping, L, A new framework for understanding vision from the perspective of the primary visual cortex. *Current Opinion in Neurobiology*, 2019. **58**: 1–10.
R28. Zhaoping, L, *Vision: Looking and seeing through our brain's information bottleneck*. arXiv preprint arXiv:2503.18804, 2025.

10 Mental disorder and stress

The hard problem of mental disorder

Chapter 9 discounted the *hard problem of consciousness* as unscientific. I think that 'mental disorder' is a related 'hard' problem but one that the science of mental mechanics can help us approach in practical terms. The key 'hard' element is semantic. What do we mean by 'mental disorder'? But, 'mental disorders' are also problematic in being complex and often comorbid. It will help us to proceed in small steps.

What about the words 'disorder' and 'pathology'? Do they apply to your problematic thoughts, feelings, and behaviours when there is no neural pathology? These thoughts, feelings, and behaviours would usually be adaptive. But they become maladaptive when they last longer or are stronger than you can tolerate – so you seek help from a therapist. In some cases (such as psychopathy) others will view you as disordered when you do not. Your failure to conform is a problem for *them* not *you*. This can get murky. In the 1960s, many were shocked that voicing opinions that were perfectly acceptable in the USA or UK could land you in a psychiatric hospital in the USSR.[1] But maybe to voice these opinions where the KGB could hear you was indeed insane?

For this, and other reasons, 'What is mental disorder?' is philosophically tricky in the same way as 'What is an emotion?' (Chapter 5). 'Mental disorder' can mean different things to different people, including diagnosticians. Problems with naming specific emotions also carry over to their specific disorders. It does not help that, even when they have purely accidental physical causes, *feelings* such as pain can seem mysterious and intractable. They are more so when we think the cause may be some form of long-term pathology rather than acute damage.

So, are aberrant mental traits more genetic/developmental/historical than event-driven? They can have multiple precursors. The problems that arise are not easy to solve.

First, if there is an underlying pathology, is it physical, mental, or both? Purely mental causes (like psychological stressors) can have neural consequences (like damage to the hippocampus). Or, is the pathology environmental

DOI: 10.4324/9781003465027-10

(e.g., being in a concentration camp, war zone, any prison, etc.). Or is it social (e.g., being looked down on as poor, 'lower class', etc.)?

Second, is 'pathology' the right word here? Where there is primary neuropathology, we may be able to answer, yes. But otherwise, it's unclear. You can see why psychiatry and clinical psychology are challenging occupations with burnout rates in the region of 30%.[R3] (Note that misdiagnosis can easily result in a dead patient with no warning.)

In this context, it is best to focus on optimal, adaptive, mental health and ignore the issue of 'pathology'. Feeling better has to be good even if you were already reasonably 'healthy'. So, in what follows, I will try and help you understand (in as much as I do myself) where unwanted, problematic mental states can come from – when they are not simply a symptom of some physical disorder. If your mental symptom is the result of a physical pathology (like liver disease), you should obviously fix the physical disorder. That is why the analgesic packets tell you to "see your doctor if headaches persist".

But before getting too analytical and academic, it may be worth reminding ourselves of the personal side. These published words are those of a patient.

Sandra's story[R9]

> When I was 10, I had what I describe as an attack. I was ... playing ... and for a moment everything was black ... I then became overwhelmed with the most horrible feeling of 'not being there'. I became 'hysterical' and ran the few blocks home. ... This began a downhill period. ... I felt as if I was 'going crazy'.
>
> (p.67) [On panic as a neural disorder see *The dynamics of disorder*.]

> The years passed and I had periods when I was relatively free of these attacks. ... But there were long periods when I was extremely frightened. I would become agoraphobic about new places and feel unable to do many things. My memories of the panic attacks are so terrible that I do not like to dwell on them even now.
>
> (p.68) [On learning and agoraphobia see *The dynamics of disorder*.]

> At twenty eight I had a major depression. ... People with agoraphobia are familiar with the feeling of being in a pit from which there is no escape. ... During that difficult time, I was for the first time given an antidepressant ... after about ten to fourteen days, I remember very vividly a weekend day when, for the first time, I felt there was a possible way out of that dark pit.
>
> (pp.70–71) [See *Stress, major depressive disorder (MDD) and posttraumatic stress disorder (PTSD)*]

> Along with my drug therapy, I began to see a counselor ... I continue to take the imipramine and Librium every day ... I gained skills to help me at anxious times and I learned a lot about panic disorder ... I realize my dark, bleak, messages are not true or are highly exaggerated ...that I will have times of crisis, but that I can use the skills I have learned and that I have medication that helps stabilize me.
>
> (pp.72–73)

Nature and nurture

Acute physical problems (accidental damage, infectious diseases, etc.) are usually attributable to environmental causes. Chronic permanent physical problems (e.g., with the lungs) can be the result of genes (nature, e.g., cystic fibrosis) or environment (nurture, e.g., smoking, asbestosis). Physical characters such as height involve both (genes for a specific height affected by, for example, level of nutrition during childhood).

So, you can be tempted to see a predisposition to any type of mental problem as arising from either a person's initial genetic nature or their later environmental nurture. But, for the bulk of mental disorders, it's best to see these interacting. The clearest exceptions are mutations of single genes that produce neural abnormalities and so behavioural consequences.

We will explore the balance between, and interactions of, nature and nurture in the next few sections looking at: interactions of genes and environment, brain development before you were born, effects of mutations and neurotoxins, and the impact of traumatic events.

Interaction of genes and environment

Many disorders have both an environmental and a genetic basis – with many genes and gene combinations making contributions. You can see this in twins. Monozygotic twins (where a single fertilized egg has split into two separate ones at the first cell division) have essentially identical genes. They are often called 'identical twins'; but most people who know them can see minor physical and mental differences. Dizygotic twins (where two separate eggs were both fertilized at the same time as each other) share only half their genes. Genetically they are as close together as other siblings but will have shared a uterine environment and usually a very similar home environment in their early years.

With ADHD (Figure 10.1), if one monozygotic twin has it, the other has a somewhat better than 50% chance of also having it. Note that with Huntington's or PKU this would be 100% and that with bipolar disorder (mania and depression, cycling) it is over 80%. So, clearly, environment has as much a role to play as genetics in ADHD in a modern society.[2] Autism, unipolar depression, and schizophrenia are similar. In all these cases the dizygotic twins have a less than 15% chance of sharing the disorder – suggesting that shared uterine environment does not play a major role compared to genetics.

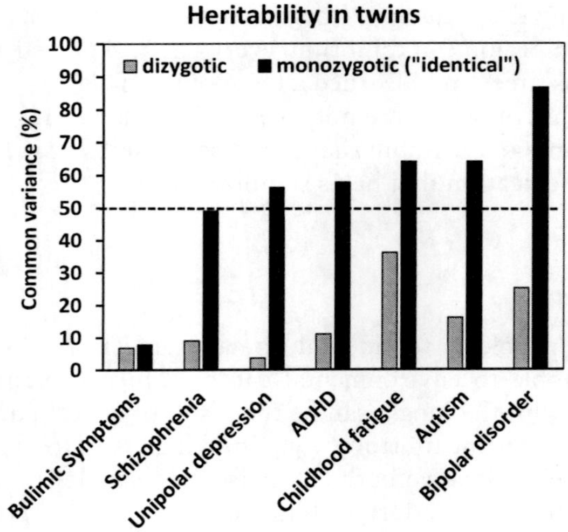

Figure 10.1 Heritability in twins. Inheritance of shared characters is estimated as the percentage of shared variance in groups of pairs of twins that are either dizygotic (dissimilar; coming from two eggs and so sharing 50% of their genes) or monozygotic ('identical', coming from one egg that split into two and so sharing 100% of their genes). Based on correlations reported in Tandon and McGuffin.[R37] ADHD = attention deficit hyperactivity disorder.

Shared uterine environment looks to make more of a contribution to childhood fatigue.

Brain development before you are born

How the brain folds itself during development is shown in Figure 10.2.

This folding trajectory offers a range of opportunities for distortion of different aspects of brain development by the uterine environment, depending on the timing of any challenge. Different conditions will involve different folding problems,[R14] for example: schizophrenia – reduced cortical folding that varies from person to person; autism – some deeper sulci with evidence for increased short- and decreased long-circuit communication; dyslexia – neuronal changes in language and auditory areas. While there is a clear genetic component in the incidence of these disorders (Figure 10.1), uterine environment is also a clear factor. For example, the anti-epileptic drug valproic acid[3] given during pregnancy results in ~10% of the children developing autism.[R34]

Effects of mutations and neurotoxins

With Huntington's Chorea, a single-gene mutation is dominant, causing production of a neurotoxic form of a protein called huntingtin. The mutant

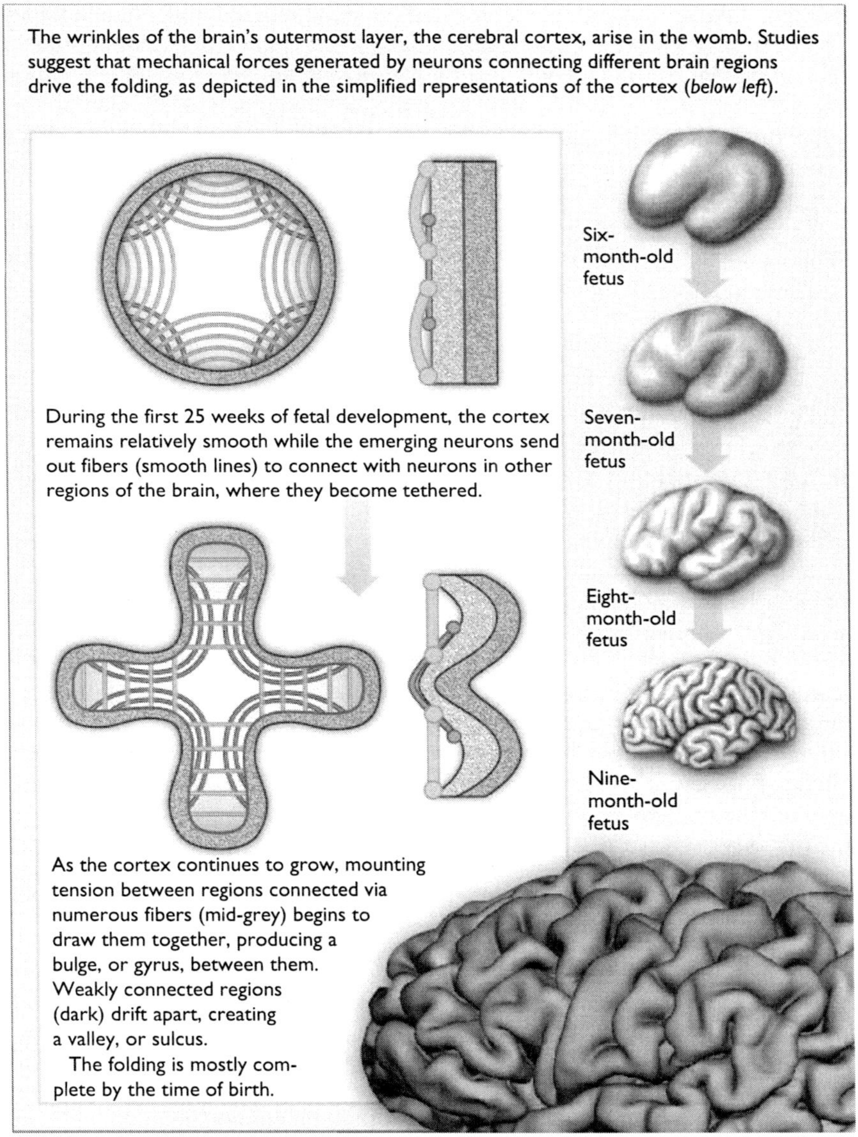

Figure 10.2 Folding the brain. From Hilgetag and Barbas[R14] with permission from Helen Barbas. Text adapted to grey scale from colour and neurons stippled.

neurotoxin immediately affects the person with the original mutation, and also affects any of their children that inherit it. About 90% of cases are inherited. Treatment can ameliorate symptoms, but death will occur within about 20 years of the appearance of initial psychiatric symptoms such as anxiety and depression and cognitive problems. Its defining symptoms are motor control problems (making it look more like a neurological than psychological disorder).

With phenylketonuria, a single-gene mutation is recessive and so both copies must be mutated to produce the primary metabolic deficit. The main effect is on blood levels of important amino acids[4]: an increase in phenylalanine and a decrease in tyrosine. If left untreated beyond the first two weeks after birth, the high levels of phenylalanine cause irreversible damage to the brain that produces a range of mental and neurological problems. You will almost certainly have had the heel-prick blood screen for high phenylalanine when you were born. If PKU is detected, the child should be immediately placed on a low phenylalanine diet that is maintained life-long.

The good news is that, provided the diet is started early and any adjustments made carefully (e.g., while monitoring phenylalanine levels), the neural and severe mental effects of having PKU can be largely prevented. I learned almost everything I know about PKU from Matt Stevenson. I first met him when he was a student in my biopsychology class. He came to me after one lecture to say that his own experience with PKU made him think that it would be relevant for the theories of behavioural inhibition that I had just been talking about. This led to me supervising his master's thesis – published as a large review, "A comparison of phenylketonuria with attention deficit hyperactivity disorder: Do markedly different aetiologies deliver common phenotypes?".[R36]

The conclusion of Matt's review was that the residual mental symptoms of well-controlled PKU were much like ADHD.

> The common deficits affect visual function, motor function, attention, working memory, planning, and inhibition. For each of PKU and ADHD separately, a subset of deficits has been attributed to a primary dysfunction of behavioural inhibition. In the case of ADHD (excluding the inattentive subtype) this has been proposed to involve a specific failure of the [Behavioural Inhibition System (BIS)[R11] for update see McNaughton and Gray[R28]]; and we suggest that this is also true of PKU. This accounts for a substantial proportion of the parallels in the superficial symptoms of both disorders and we see this as linked to prefrontal, rather than more general, dysfunction of the BIS.
>
> (Stevenson[R36, p. 63])

Matt went on to obtain a PhD at the University of Copenhagen. Then, after several post-doc positions at different hospitals across the city, he returned to

New Zealand with his family to practice as a hospital psychologist. He is also still involved in publishing research. The dietary restrictions needed to control PKU are very burdensome; but they have allowed Matt to live a successful life.

As Matt noted in his review, the primary causes and development of ADHD for details, see Stevenson[R36] are quite different from PKU. Many complex genes appear to be involved in ADHD and to interact with environmental factors.

> The biochemical and neural pathologies of PKU and ADHD are quite distinct in their causes and detail; but they result in the disorder in the brain of large amino acid levels, dopamine and white matter that are very similar and could explain the overlap of symptoms within and between the PKU and ADHD spectra.
>
> (Stevenson[R36, p. 63])

Huntington's, PKU, and ADHD produce psychiatric symptoms. But the origin of these is mostly neurological with clear brain changes producing the mental symptoms. In turn, the brain changes are produced by neurotoxic compounds resulting from single gene (Huntington's, PKU) or multiple gene (ADHD) mutations. In the case of ADHD there is a strong additional role for environmental factors.

The impact of traumatic events

Finally, let us look at a clear case where brief environmental events (or more chronic exposure in war veterans) result in long-term post-traumatic stress disorder (PTSD). As an example, in ~37,000 responders to the 11 September 2001 attack on the World Trade Centre that killed nearly 3,000 people, there was an incidence of PTSD of 15% (and similarly for depression) in 2015–2016.

Frank Mann and colleagues[R23] looked at 20-year PTSD symptom trajectories in ~13,000 of these responders.

> Symptoms were stable in the short term but changed significantly over two decades, peaking over a decade after exposure and declining modestly thereafter. ... The median time before symptoms improved was 8–10 years for PTSD cases .... Most experienced improvement after a decade but approximately 10% reported elevated symptoms two decades after trauma.
>
> (Frank Mann and colleagues[R23, abstract])

Even here, with a clear causal event and an extremely long-term stable change in mental state, we cannot rule out nature from the equation.

> PTSD is unique among adult mental disorders because exposure to a traumatic event(s) is required to meet criteria for a clinical

> diagnosis. Developmental capacities and conditions of early childhood may increase both risk of trauma exposure and of PTSD. Rather than being solely a proximal response to trauma, the roots of PTSD involve predisposing developmental origins. ... PTSD in adulthood rarely, if ever, develops in anything but the context of other disorders. ... Both severe [childhood] maltreatment and prior history of mental disorders were found to independently predict adult PTSD. Note, the experience of mild trauma during childhood was not found to increase the risk of PTSD by the adult years, providing little evidence for the hypothesis of stress-sensitisation from moderate trauma.
>
> (Poulton et al.[R31, p. 3])

Nature and nurture: an overview

Taking the previous sections together, it is clear that what drives some mental problems are more neurological factors and what drives others are more psychological (Table 10.1). But, as we will see shortly, environmental and dynamic factors can be important in relation to more neurological causes (* in Table 10.1) while neurological (particularly personality) factors can be important in relation to more psychological causes (# in Table 10.1). Psychologically-induced stress can also result in neurological changes.

The dynamics of disorder

Disorder can also involve dynamic changes in neural systems, interacting with the environment. As an example, panic is fundamentally simple, but its causes can be very complicated.

Panic, as an adaptive response, is most simply elicited via suffocation or other situations perceived as deadly. It also occurs as a symptom of high levels of adaptive anxiety, resulting in a catastrophic interpretation of high

Table 10.1 Variation among disorders in neural versus psychological cause

Neurological bias	*Mixture#*	*Psychological bias#*
Huntington's Chorea	ADHD	"Stress Disorder"
Phenylketonuria+	Separation Anxiety	Endogenous Depression
Parkinson's Disease	PTSD	Reactive Depression
Alzheimers Dementia	Panic/Agoraphobia	
Schizophrenia*	Social Anxiety	
Bipolar Disorder*	Obsessive Compulsive Disorder	
Panic ("Irritable Heart")	Simple Phobia	
Autism		

+ Major effects of diet.
* Environmental and dynamic factors (esp, stress) can exacerbate/ ameliorate symptoms, etc.
# Personality (=neural) factors can affect severity.

arousal levels. In these cases, neither it nor its causes would be the result of a mental disorder – just of extreme situations.

Panic can also occur as an essentially adaptive symptom to maladaptive anxiety (Figure 10.3A). Excessive reactivity of areas like the amygdala and hippocampus (e.g., to anticipation of a threat) can generate maladaptive anxiety that, because it increases arousal, triggers a panic attack.

Alternatively, panic itself can be maladaptive. For no apparent reason, even in the safety of the clinic, you feel "An abrupt surge of intense fear or intense discomfort that reaches a peak within minutes" with symptoms that can involve the heart, breathing, chest pain, dizziness, fear of 'going crazy', or dying.[5] This can result from hyperactivity of the periaqueductal grey (which would mediate adaptive suffocation alarm) and in someone with a stable personality would be seen as closer to 'irritable heart syndrome' than 'panic disorder'".[R17] Or it can result from hyper-reactivity of the PAG where, in agoraphobia and panic disorder, the "panic attacks result from the catastrophic misinterpretation of anxious symptoms".[R18, p. 253] However, particularly in those with a neurotic personality, a panic attack can immediately elicit fear (left side of Figure 10.3B). Learned anticipation of this fear can generate anxiety (particularly agoraphobia; right side of Figure 10.3B). Resultant arousal can then increase the incidence of panic attacks through much the same mechanisms as panic induced by injections of adrenaline. This in turn can increase anxiety via learning and also (through a mechanism called kindling) generate neural disorder in the anxiety systems (Figure 10.3C).

The end result in Figure 10.3C is true comorbidity where both panic/fear and anxiety systems are dysfunctional and potentially mutually reinforcing. "Yet more complex are syndromal comorbidities of anxiety that go beyond the two dimensional fear/anxiety systems: depression, substance use disorder, and attention-deficit/hyperactivity disorder".[R24, p. 1053] Additional complexity is added to the bulk of psychiatric disorder by the stress response.

What is stress?

Here is another of those 'What do you mean by ...?' questions. We often talk about someone being exposed to stress. But it's better to say that they are exposed to a stressor and then may experience a stress reaction internally. "Stress" as a simple word could be either or both of these. I find it easiest to define the stress response by the release of the hormone cortisol (CORT[6]) into the bloodstream.

We also often think of a stress response as automatically a bad thing. But *brief* periods of exercise, or *brief* immersion in cold water (as in the Nordic cold plunge) can elicit a good form of stress reaction ('eustress'). This mobilizes energy, inhibits inflammation, increases arousal and (together with endorphins) enhances performance. These positive stressors (and the negative ones, illness and injury) are all simple physical stimuli. So, you can get really tangled if you see 'stress' as being a fundamentally psychological state.

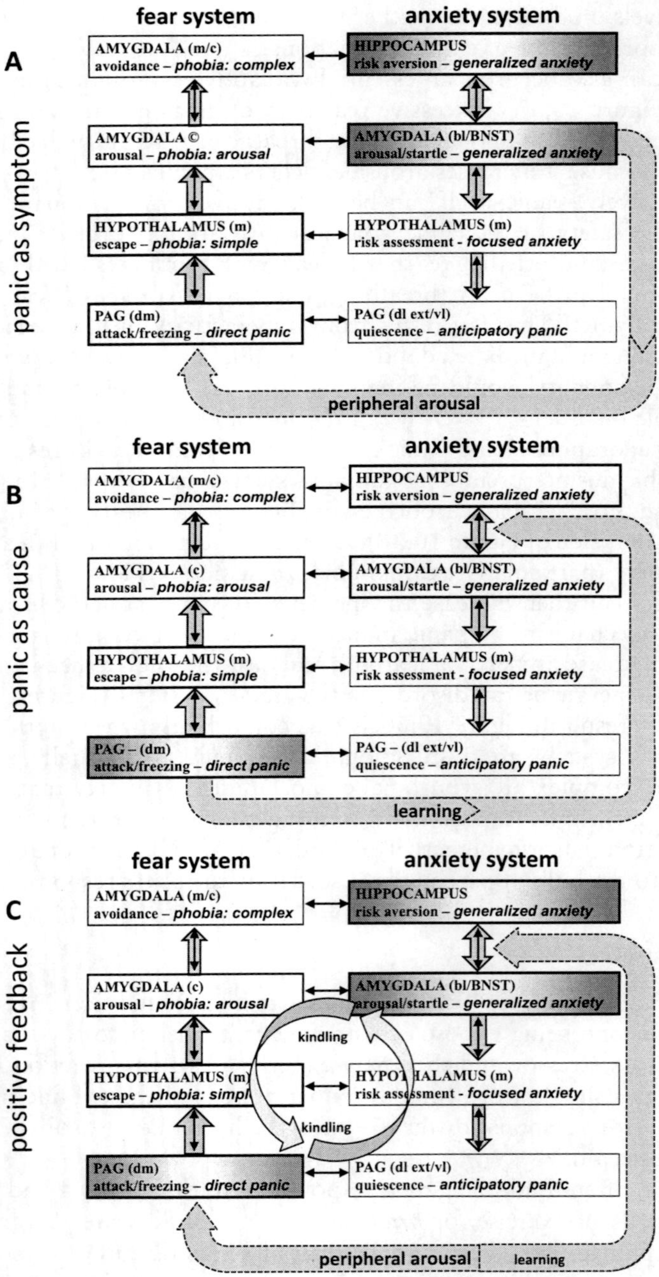

Figure 10.3 Disorder dynamics. (A) Panics as symptom. (B) Panic as cause. (C) Positive feedback. For details see text. See also 'Fear versus anxiety'. Abbreviations: c = central; BNST = bed nucleus of the stria terminalis; bl = basolateral; dl ext = dorsolateral external; dm = dorsomedial; m = medial; vl = ventrolateral. Updated from McNaughton and Corr[R26] with permission of Neil McNaughton. For details see text.

Stressors can be purely psychological. The World Trade Centre responders whose PTSD we discussed earlier had not experienced physical damage but reacted to their memory of the experience. Similarly, in parents of children diagnosed with leukaemia, anticipatory grief produces higher levels of stress hormones than anticipatory denial – with a reversal of these relative levels after the death of the child. Many parents still have high levels 2 years after the death of the child. One couple with originally high levels, reverted to low ones after learning that their child was misdiagnosed (and so survived).[R15,16]

In addition to CORT, the stress response involves: (a) release into the blood of a range of other hormones (including adrenaline) from the pituitary-adrenal system; (b) release onto organs of adrenaline and noradrenaline by the sympathetic nervous system; and (c) diffuse release of ascending monoamines (serotonin, noradrenaline, dopamine, and histamine) into the forebrain. The next section covers this in more detail. Skip it if you want to keep things simple. The forebrain systems are also important for various mental disorders and we will cover this later.

**The stress response and its control*

A sketch of 'the stress response' is shown in Figure 10.4. This is simplified in that, for example, it does not show that the hypothalamus releases corticotropin-releasing hormone which causes the pituitary to release adrenocorticotropic hormone which causes the adrenal glands to release CORT. Importantly both of these hormones can have their own actions as well as being part of the CORT-release cascade.

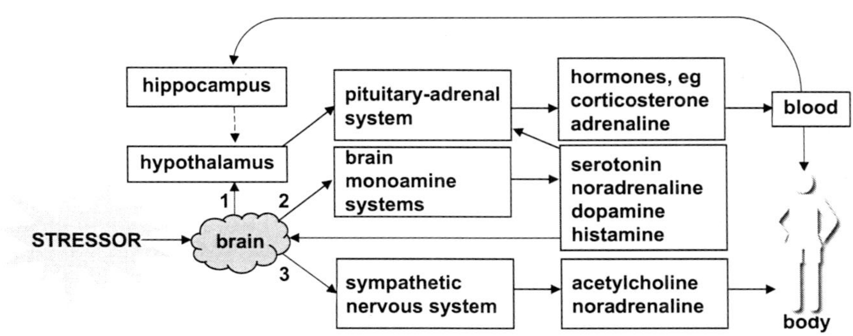

Figure 10.4 The stress response. A physical or psychological STRESSOR in the environment activates complex cascades of events. (1) Brain circuits, via the hypothalamus and pituitary-adrenal system, release of hormones into the blood to act anywhere there are appropriate receptors. In the case of CORT, this causes feedback inhibition (dashed arrow) when detected by the hippocampus. (2) Activation of brainstem monoamine systems causes release of serotonin, noradrenaline, dopamine, and histamine into the forebrain, including the prefrontal cortex. (3) The sympathetic nervous system releases acetylcholine and noradrenaline onto a range of specific target organs (see Figure 5.3).

For our purposes, however, we can see stress as simply releasing a soup of hormones (or, in the brain, monoamines that have diffuse effects rather like hormones). Together with activation of the sympathetic nervous system, these chemicals produce short-term adaptive changes when the body needs to be prepared to cope with tissue damage or metabolic problems. Importantly, a change in state is needed because it would be maladaptive to remain in an activated state too long.[7]

Release of CORT is normally controlled by a feedback loop via the hippocampus (Figure 10.3). This has one of the highest concentrations of CORT (and other endocrine) receptors in the brain.[R21] When the hippocampus detects CORT in the bloodstream, it sends an inhibitory signal to the hypothalamus to dampen down the pituitary-adrenal cascade. But stress in general and CORT in particular also affect the hippocampus.[R35, pp. 215–225] With low-level, acute, stress hippocampal connections are increased from lower numbers to higher numbers in males (but from higher numbers to lower numbers in females).[R4]

But at higher chronic levels of CORT hippocampal function is impaired and its cells can even be killed. This has the potential to produce a vicious cycle of increased CORT release and increased hippocampal damage. This may be an important part of the difficulty in treating some psychiatric disorders.

Stress, major depressive disorder (MDD) and posttraumatic stress disorder (PTSD)

PTSD and major depression are linked to high CORT levels. They can produce a decrease in hippocampal volume that is related to the severity of trauma and the duration of the disorder, respectively.[R35, p. 221] Since the hippocampus produces feedback inhibition of CORT release, this may explain why these disorders have chronically high CORT levels and are so difficult to treat.

> Why should the stress-induced activation of the HPA axis, a biological system that is life saving and enables us to fight or escape our enemy, lead to such a bad thing as depression? The answer, from an evolutionary point of view, is that depression – if you are a fawn in a cold barren land, or a defeated gorilla that has fallen in the dominance hierarchy – is an adaptive response. Depression stops you dispersing energy in the pursuing of unavailable goals, prevents further aggressive behaviour from the dominant animals, and signals your difficulty.
>
> (Pariante[R30])

The same is true of separation distress in humans and other mammals. Interestingly, in bonnet macaques (who live in large social groups) separation from the mother produces anxious behaviour and interaction with other adults, resulting in help, and sometimes even adoption. In pigtail macaques (who live in small groups) it rapidly produces a facial expression of grief and a severe

> depressed stage, with marked curtailment of movement, the rolled up body with oft-hidden face and the general reduction of interaction with others would appear to conserve energy and to reduce the likelihood off attack, i.e. to sustain and prolong life until relief might be obtained.
>
> (Kaufman, IC & Rosenblum[R19, p. 693])

So, separation anxiety, followed by depression when separation is prolonged, is normal and adaptive. You will likely have experienced them (but not remember this), peaking at about 1 year old. But separation anxiety (like other anxiety) can be classified as a disorder if it is excessive or occurs later in development – particularly in adults – with its release of CORT increasing the chances of co-morbid depression.

Anxious agitation and depressed inactivity are two contrasting ways (active and passive) of attempting to avoid metabolic and external threat. This explains the physical bias both of the typical features of MDD (loss of interest/pleasure, fatigue, poor concentration, feeling worthless or excessively guilty, recurrent thoughts of death) and of the opposing active/passive extremes of some of them in different cases (weight loss/gain, insomnia/hypersomnia, psychomotor agitation/retardation).

If we see MDD as resulting from excessively extended stress and PTSD as resulting from brief but excessively strong stress, it is not surprising that they are often comorbid and share many symptoms. A key difference is that "the depressive symptoms of PTSD patients are closely related to the intensity, re-experience and avoidance of traumatic events, while the depressive symptoms of MDD are less directly related to prior traumatic events".[R39, p. 713] They show similar patterns of increased task-evoked brain activation relative to healthy controls with similar activation to each other in the inferior frontal gyrus with stronger results in some areas with PTSD and other areas with MDD.[R39] Importantly, the development of disorder in both cases depends not only on environmental events but also on the dependence on genetic and developmental factors (see Nature and nurture – events) of their interpretation. This depends, among other things, on predisposing personality.

Neuroticism – a key personality factor for internalizing disorders

Neuroticism is one of the five[8] fundamental factors of the 'Big Five' model of personality.[R24, 8, Fig. 1, p. 56] So, even with high scores it is not a disorder. However, it appears to be a risk and maintaining factor for a range of internalizing disorders (where the patient's focus is internal) as opposed to externalizing disorders (where the focus is outward, generating disruptive external behaviour).

> Several theories have posited a common internalizing factor to help account for the relationship between mood and anxiety disorders. These disorders are often co-morbid and strongly covary. Other theories and data suggest that personality traits may account, at least in part, for co-morbidity between depression and anxiety. ... [high] neuroticism may be the core of internalizing psychopathology.
>
> (Griffith et al.[R13, p. 1125])

Neuroticism, broadly, is a tendency to negative evaluations and emotions. It results in emotional instability, poor responses to stressors, and can result in substance abuse in attempts to reduce unpleasant feelings. Its interaction with a stressful life event often triggers an episode of anxiety or depression. Importantly, neuroticism may be relatively more important than the stressor. With an Australian bush fire that produced chronic PTSD in a third of the firefighters, neuroticism predicted the PTSD better than the extent of exposure to the fire.[R25] But as we noted in the section on *the impact of traumatic events* earlier highly stressful events can produce long-term changes in personality that can be interpreted as increased neuroticism. To some extent the goal of therapy is to reduce neuroticism.

Pharmacology of the neurotic disorders

A range of superficially distinct disorders are classified as 'neurotic' in the *ICD-10* classification system.[9] These include major depressive disorder, dysthymia, post-traumatic stress disorder, generalized anxiety disorder, social anxiety disorder, specific phobia, panic disorder, agoraphobia, and obsessive-compulsive disorder.[R40] In all of these disorders you have an excessive internal response to threat (experienced or anticipated). But, across the disorders, your symptoms and their biology are different.

The best evidence for differences among the neurotic disorders is conventional drug treatment. A wide range of different classes of chemicals have been used to treat them. Variation in their effects with diagnosis is shown in Table 10.2. You can ignore both the pharmacological and diagnostic details but see from the shaded intensity of the pattern of effects that not all drugs are effective for all disorders.

Of particular note, the drugs in the bottom four rows of the table share (as a group) effects on generalized and social anxiety disorder and share a lack of effects on panic disorder and obsessive-compulsive disorder. These drugs are chemically quite different from each other and have common neural actions through quite distinct pharmacological pathways. Similarly, some drugs that are effective for panic disorder are relatively ineffective in obsessive-compulsive disorder. That said, our mental mechanics also need to explain

Table 10.2 Comparison of the effects of different drug classes on internalizing disorders. Focus on the shaded pattern more than the details. Drugs are ordered in approximate relation to general effectiveness from most effective to least effective.

	GAD	*SAD*	*MDD*	*PD*	*PTSD*	*OCD*	*SP*
Ketamine	+	+	+	+	+	+	
Mirtazapine	+	+	+	+	+	+	
SSRI	+	+	+	+	+	++	(+)
SNRI	+	+	+	+	+	+	
TCA_2 (≈CMI)	+	(+)	+	++	(+)	++	
TCA_1 (≈IMI)	+	(+)	+	+	+	(+)	0
MAOIs		+	+	+	(+)	(+)	(+)
NRI		(+)	+	(+)	(+)		
Buspirone	+	(+)	+	0	0	(+)	
BDZ_2	+	(+)	(+)	(+)	0	0	0
BDZ_1	+	(+)	0	0	0	0	0
Pregabalin	+	+	0	(0)			

Disorder abbreviations: GAD = Generalized anxiety disorder; MDD = Major Depressive Episode; OCD = Obsessive-Compulsive Disorder; PD = Panic Disorder; PTSD = Post-Traumatic Stress Disorder; SAD = Social Anxiety Disorder (previously Social Phobia); SP = Specific Phobia.
Activity codes: blank = no data; 0 = inactive; (0) equivocal inactivity; (+) = equivocal or high dose activity; + = active; ++ = more strongly active than related drugs.

BDZ_1 early benzodiazepines, e.g., chlordiazepoxide and diazepam are administered at typical (i.e. low) antianxiety doses. Other sedative antianxiety drugs (barbiturates, meprobamate) have similar effects.
BDZ_2 later high-potency benzodiazepines, e.g., alprazolam. The antipanic effect is achieved at higher doses and this has also been reported with equivalent high doses for BDZ_1.[R29] See also van Marwijk et al.[R38]
MAOI Monoamine Oxidase Inhibitors, e.g., phenelzine.
SSRI Selective Serotonin Reuptake Inhibitors, e.g., fluoxetine, citalopram.
TCA_2 high-effect tricyclics, e.g., clomipramine.
TCA_1 Imipramine and related tricyclic antidepressants, but excluding clomipramine.

Table and legend adapted from McNaughton and Glue[R27] under the terms of the Creative Commons Attribution licence (http://creativecommons.org/licenses/by/4.0/).

how drugs like serotonin reuptake inhibitors can improve at least some cases of all the neurotic disorders albeit slowly; while drugs like ketamine can quickly improve cases resistant to conventional treatment (drugs or psychotherapy). The next two sections discuss how fear/panic, anxiety, and obsession each appear to depend on distinct neural systems.

Anxiety versus fear

In Figure 10.3 subcortical neural systems controlling panic/fear and anxiety are shown as distinct. But, in both cases, they control responses to dangers that can be learned or innate (e.g., snakes, Figure 5.5).

The separation of the fear from the anxiety system is based primarily on the lack of action of 'anxiolytic' drugs on panic and phobic avoidance. Anxiolytics have been used to define the neural elements of the 'anxiety system' and link them to approach-avoidance goal conflict.[R28] The anxiety system uses inhibition of approach to achieve safety. But it also generates risk assessment and can allow one to cautiously move *toward* less severe dangers. Clinical anxiety often involves an exaggerated assessment of risk and so excessive inhibition of behaviour.

The fear (fight/flight/freeze) system is directionally opposite. Its structures function to move one *away* from danger and were originally identified by electrical stimulation eliciting defensive behaviour. Its lowest level generates panic when danger is severe and immediate (especially with death likely). Higher neural levels deal with escape and then avoidance when there is progressively more time for action (Figure 10.3).

I should emphasize that this is very much my own perspective as detailed in McNaughton and Gray[R28] Even a recent psychiatric diagnostic system (*DSM-5*[R1]) is confused about the fear/anxiety distinction with 'anxiety disorders' including 'fear':

> Anxiety disorders include disorders that share features of excessive fear and anxiety and related behavioural disturbances. *Fear* is the emotional response to real or perceived imminent threat, whereas *anxiety* is anticipation of future threat. Obviously these states overlap. ... *Panic attacks* feature prominently within the anxiety disorders as a particular type of fear response. Panic attacks are not limited to anxiety disorders but rather can be seen in other mental disorders as well.
>
> (American Psychiatric Association[R1, p. 826])

Specific phobias can usually be treated with psychological therapy (particularly Cognitive Behavioural Therapy); while panic and avoidance can usually be treated with either psychological or pharmacological therapy (panicolytic and antidepressant but not anxiolytic drugs). Generalized anxiety and, to a lesser extent, social anxiety can be treated with psychological or pharmacological therapy (Table 10.2). However, in all these cases, improvement tends to take months, is ineffective in as many as a third of patients and, with drugs, can require maintenance. That is, the drugs reduce symptoms more than curing the disorder (as with aspirin for a headache).

Anxiety versus obsession

The remaining outlier is obsessive-compulsive disorder (OCD). In earlier versions of the *DSM*, it was classified as an anxiety disorder. But *DSM-5*[R1] put it in a separate category. *Obsessions*, here are

1. Recurrent and persistent thoughts, urges or images that are experienced, at some time during the disturbance, as intrusive, unwanted,

> and that in most individuals cause marked anxiety or distress. [AND] 2. The individual attempts to ignore or suppress such thoughts, urges, or images, or to neutralize them with some thought or action (i.e., by performing a compulsion).[10]

Compulsions are repetitive behaviours or thoughts driven by the obsession and designed to deal with its consequences. To be classified as disorder (as usual with *DSM*), the obsessions and compulsions must be excessive or inappropriate. A key point is that execution of the compulsion reduces distress but failure of execution can produce anxiety as a symptom (much the same as panic can).

The nature of the compulsions in OCD is interesting.[R32] Excessive *self*-cleaning rituals occur in 85% of cases (decontamination rituals in ~20%, harm prevention in ~15%, *object* cleaning in ~5%), repeating rituals (e.g., going in and out of doors) in ~50%, checking (doors, locks, etc. – much more than you ever would normally) in ~50%, and others. The sum is greater than 100% because many cases have multiple compulsions.

The key point for a behaviourist is that all of these are avoidance behaviours. But, unlike the avoidance controlled by the amygdala (Figure 10.3), the dangers are all undetectable. Infectious micro-organisms are a prime example. So, unlike simple avoidance, there can be no signal that safety has been achieved. There are innumerable notices telling us to wash our hands (especially with COVID around). But when do you stop? The notices typically say 20s wash + 20s dry (the drying bit is clearly important since the bugs like to be in water). But, with OCD, you will rinse your hands much longer; and you may count the exact number of hand-wrings each time.

So, we have a simple sequence. Detection of danger (often normal). Initiation of escape/avoidance behaviour (often normal). Lack of safety signal. Ritualized (abnormal) extension of avoidance. Safety. Repeat with more extension occurring over time until avoidance is blocked by life logistics. Dysfunctional other avoidance, dysfunctional anxiety, etc.

We just went off the top of Figure 10.3. That has subcortical systems dealing with fixed events in a simple fashion. The key locations for obsessions are the anterior cingulate cortex (one step up from the amygdala) and then caudal medial prefrontal and orbital frontal cortex.[R12] These are all involved in prediction and checking for danger. The key location for compulsions is the basal ganglia[R12,32] and their connections to, for example, anterior cingulate cortex.

> For both functional and neurological reasons, then, we arrive at the following view of OCD. Obsessions and compulsions can arise from overactivity in the cingulatebasal ganglia circuitry. ... Often the overactivity will simply give rise to avoidance behaviour (or successful checking), will not produce major increases in anxiety, and will not lead to the seeking of clinical help. ... Where the frequency of fully

> fledged avoidance behaviour is very high, or the adequate stimuli elicit avoidance in problematic situations, or the avoidance response is in some other way blocked, then there will be conflict between this response and others ... with consequent anxiety. Such anxiety should be most likely (or will occur at a lower threshold) in those with a neurotic introvert personality. This is one scenario, given the normal progression of symptoms. [There is] an alternative scenario: pathologically maintained anxiety (most likely generalized anxiety) could provide an unusually high level of input to the cingulatebasal ganglia circuitry .... This input could then trigger any latent obsessions and/or compulsions.
>
> (McNaughton and Gray[R28, pp. 479–480])

Obsession/compulsions, like panic, may then be generated by specific neurological hyperactivities that produce reportable symptoms but no major distress and so no clinical complaint. But when the result is perception of unavoidable danger, and the person has a neurotic personality, distress and anxiety are generated and amplify the original abnormality. Conversely, abnormal levels of anxiety (linked to neuroticism) can trigger panic or obsession, especially these were already over-sensitive.

Ketamine and neuroticism: a double-hit hypothesis of internalizing disorders[R27]

Why are internalizing disorders so interlinked? How do they share as a risk and maintaining factor? Why is ketamine an effective treatment in cases resistant to conventional drugs?

The basic idea is simple (Figure 10.5). If you are only neurotic or only have high trait anxiety (panic/obsession/depression) you will not have disorder. But these factors amplify each other – a 'double hit' that can generate disorder. The specific disorder traits reflect sensitivities to your reactions to chronic or strong acute stress. Anxiolytic drugs' impact on specific anxiety-related triggers. Neuroticism affects negative, particularly threatening, evaluations in general. The controls of neuroticism and of the specific traits are complex with genetics, epigenetics, and development environment all playing a part.

Our key speculative hypothesis is that conventional treatments act on the specific disorder systems described above (e.g., anterior cingulate for OCD). But ketamine acts to alter neuroticism and so (usually) improves all the disorders. This also appears true of anorexia.[11] Specific serotonin reuptake inhibitors (like Prozac) also have the capacity to improve all these disorders by increasing the personality factor of stability.[R34] The effects of SSRIs on anorexia are more complicated.[12] But serotonin is not specific to aversive events[R35] and has weak, slow effects.

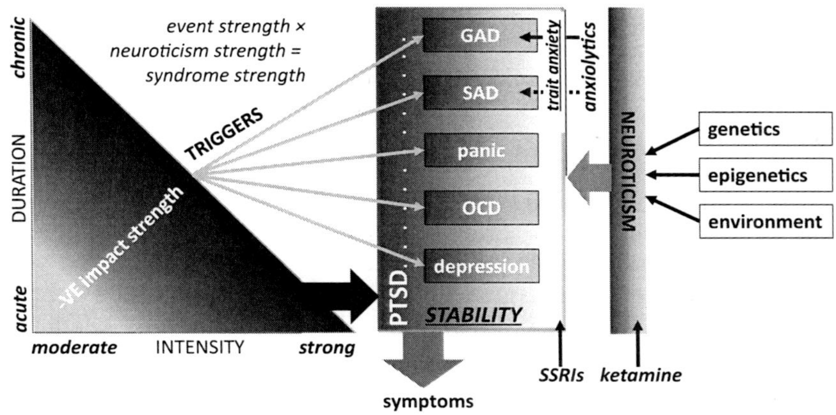

Figure 10.5 Diagrammatic representation of the 2-hit hypothesis for disorders known to be affected by ketamine. Disorder is held to result when high neuroticism (resulting from genetic, epigenetic, and prior environmental factors) is combined with a high level of a specific trait linked to GAD, SAD, panic, OCD, or depression. High levels of these specific traits can be triggered by moderate chronic or strong acute stress. In the latter case PTSD may result. Specific anxiolytic drugs affect the specific trait linked to GAD and to a lesser extent SAD (see Table 10.2). SSRIs act on a background higher order trait of 'stability',[R7] which would not be specific to aversive events[R5] and which would slowly alter the specific trait levels. Ketamine acts to affect neuroticism and so alters the disordered expression of all the specific traits. Figure recoloured and legend adapted from McNaughton and Glue[R27] under the terms of the Creative Commons Attribution licence (http://creativecommons.org/licenses/by/4.0/).

Schizophrenia – a different 'double' hit

We have already briefly looked at the, early, likely-neurodevelopmental origin of schizophrenia, particularly cortical folding (Figure 10.2). We also described features of the, developmentally later, 'broken mind'; and suggested that negative symptoms come from neuronal loss in key structures, while positive symptoms come from neural hyper-reactivity.[13] This sequence implies a double hit, but one unlike the neuroticism story.

The earlier, neurodevelopmental, hit is complex. There can be many causes of the brain's distortion: pre- and peri-natal (infection, famine, placental pathology, low birth weight) and early developmental (urban environment, childhood trauma, minority/immigrant status).[R33] But it's only a first hit, young children do not have full-blown schizophrenia.

This first neurodevelopmental hit is also not absolute. We already noted that 'identical' twins show only 50% concordance for schizophrenia (Figure 10.1). The brain of a healthy twin starts abnormal but normalizes by late adolescence.[R33] There are clearly environmental factors at work.

As a result, we must be prepared for considerable variation in the extent and nature of schizophrenia. It is worth remembering, here, that PKU and ADHD can produce very similar symptom patterns through genetically and mechanistically distinct paths.[14] The term 'schizophrenia' may label a range of equally distinct types.

The second hit is no less complex. For many schizophrenics, adolescence appears to trigger the appearance of the disorder (a bit earlier in boys than girls). But it can also appear earlier. Both early and late forms are linked to neuronal loss in temporal and prefrontal cortex.

Interestingly, there is an argument for at least some 'negative' symptoms being due to high rather than low brain activity (Figure 10.6). A failure of inhibitory function in the hippocampus makes it hyperactive. This scrambles its normal functional rhythmicity, disrupting not only the hippocampus but also the medial prefrontal cortex and the amygdala. Even without

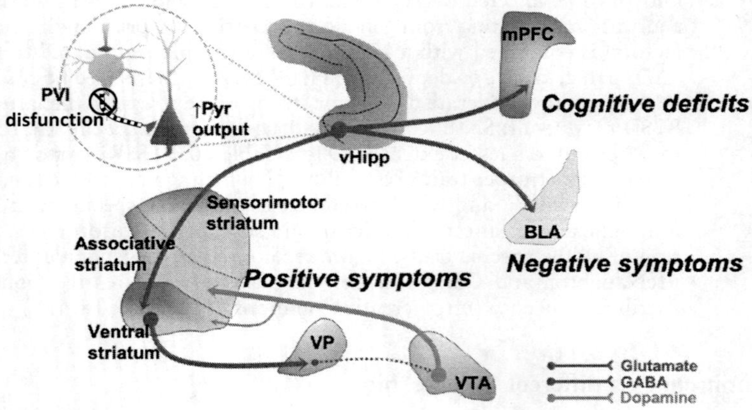

Figure 10.6 The hippocampus and positive and negative symptoms of schizophrenia. The ventral anterior limbic hippocampus in humans (vHipp) is proposed to be hyperactive and dysrhythmic in schizophrenia due to decreased parvalbumin interneuron (PVI) inhibition of pyramidal (Pyr) neurons. This is thought to lead, through a ventral striatum-ventral pallidum (VP) pathway, to an overdrive in the activity of ventral tegmental area (VTA) dopamine neurons that project to the associative striatum. The resulting striatal hyperdopaminergic state has been linked to the positive symptoms of schizophrenia. Additionally, a hyperactive hippocampus can also interfere with the function of other circuits. For instance, disruption of medial prefrontal cortex (mPFC) and basolateral amygdala (BLA) could potentially lead to cognitive deficits and interfere with emotional responses, leading to negative symptoms, respectively. Therefore, a hyperactive dysrhythmic limbic hippocampus potentially disrupts multiple circuits and could contribute to the three main symptom clusters of schizophrenia. Figure recoloured and legend adapted from Gomes and Grace[R10] under the terms of the Creative Commons Attribution licence (http://creativecommons.org/licenses/by/4.0/).

cell loss, this would produce negative (cognitive and emotional) symptoms. Hippocampal hyperactivity could also (by inhibition of inhibition) produce excessive activation of dopamine systems (ventral striatum➡ventral pallidum➡ventral tegmental area), generating positive symptoms.

The current antipsychotic drugs are dopamine antagonists and so, on this scheme, would affect only positive symptoms. But hyperactivity in the ventral hippocampus could be a novel target for treatment of at least some of the negative symptoms.[R10] In my view such hyperactivity could also be the cause of neuronal loss in hippocampal, prefrontal, and other areas – which are known for excitotoxicity. It could, therefore, even be possible to target this second hit and prevent occurrence of this damage (although not reverse it once it had occurred).

As with internalizing disorders, stress makes important contributions to schizophrenia (which is neither internalizing nor externalizing). One way it can do this is via the release of dopamine by CORT (Figure 10.4), which would activate positive symptoms. Also remember that high CORT can produce hippocampal damage.[15] Although not the sole trigger, stressful life events can trigger the occurrence of schizophrenia (SZ) and often trigger relapse.

> SZ may be a progressive neurodevelopmental disorder in which early pathologic factors, such as genetic vulnerability, prenatal insults, and obstetric complications, create an abnormal neurobiological substrate that is more vulnerable to the development of SZ symptoms. Events that may trigger actual onset of SZ include both (1) maturational processes, such as programmed cell death and synaptic remodeling, and (2) postnatal environmental insults, such as psychosocial stress, TBI, and perhaps even substance abuse. A mechanism by which development unmasks latent vulnerability is plausible.
>
> (Corcoran et al.[R6, p. 37])

It has been suggested that, rather than novel treatments,

> a more effective approach would be to prevent the transition to SZ in susceptible individuals. A focus on stress, which has been shown to be a predisposing factor in risk for SZ, is a possible avenue that has shown promise in preclinical studies.
>
> (Gomes and Grace[R10, p. 1])

The same may be true of other neurodevelopmental disorders, all of which may be better termed spectra.

Neurodevelopmental spectra

This picture of schizophrenia applies, with surprisingly little change, to "other childhood-onset conditions, including attention deficit hyperactivity disorder, intellectual deficiency, autism spectrum disorders (ASD) and epilepsy. Optimally, broader (less specific) risk could lead to translational research enabling broad preventive measures".[R33, p. 1228]

As we noted with cortical folding (Figure 10.2), the neurodevelopmental problem may depend on the timing of pre- or peri-natal insult (e.g., infection) interacting with a common liability to abnormality across the resultant disorders. Such timing and liability differences will not have sharp boundaries, allowing spectral variation both in the nature and dysfunctionality of the disorders. So, schizophrenia, ADHD, and intellectual deficiency should likely all be labelled as spectra (as with autism).

Such spectra can be viewed as reflecting personality traits and so could include individuals with scale scores well below the clinical range. With intellectual deficiency, one might argue that there is no such thing as having too much intellect. Likewise,

> ADHD is best viewed as the quantitative extreme of genetic and environmental factors operating dimensionally throughout the distribution of ADHD symptoms, indicating that the same etiologic factors are involved in the full range of symptoms of inattention, hyperactivity and impulsivity.
>
> (Larsson et al.[R20, p.73])[16]

That said, it has been argued that people with even moderate schizotypy can "display poorer psychosocial functioning, lower rates of marriage, increased use of psychiatric medications, and increased utilization of psychiatric services" while also generally being at risk of developing schizophrenia.[R22, p.26] Similarly, what used to be referred to as Asperger Syndrome and seen as benign is now simply included in autism spectrum disorders and seen as requiring help and treatment.

Should we talk about 'mental disorder'

First, let us look at 'disorder'. This is usually taken to equate with pathology – the breakdown of a system.

But the spectra (and the nature of the neurotic disorders) explain why both the *DSM* and *ICD* diagnostic systems emphasize that the symptoms must be problematic (for the patient or significant others or an autocratic government) *and* prolonged for something to be classified as a disorder. With the neurotic disorders, there is hyperactivity of systems where even extreme symptoms can be adaptive (as responses to truly extreme situations).

With the neurodevelopmental disorders it is clear that low levels of the typical disorder symptoms can just be normal variation even if not obviously adaptive (in comparison to obviously maladaptive high levels). That said, with intelligence, schizotypy and the ADHD symptom package (and possibly even autism) we can argue that there is a bell curve of distribution of the strength of the relevant characters in our population because having too low a level will be equally problematic to having too high a level.

Next let us look at 'mental'. Certainly (Table 10.1), some disorders have a stronger psychological source (reactive depression as an extreme) and others a more neurological one (Huntington's chorea as the extreme). But in all cases, nature and nurture are intertwined. Likewise, a pure psychological stressor can result in hippocampal damage and a single gene mutation[17] produce a spectrum of psychological changes.

Overall, it's best to see mental mechanics as explaining all psychology and behaviour. Then, of course, it does not matter whether what bothers you is 'normal' or 'abnormal', you need to do something about it. If you look at yourself as both a biopsychologist and an experimental participant you should be able to work out the mental mechanics of your problem via self-experiments. This may help you apply a biopsychological solution to yourself or, in the case of your neighbour's loud music disturbing your sleep, give you biopsychological reasons for taking action directed to others.

Notes

1 For a long article on this, see https://en.wikipedia.org/wiki/Political_abuse_of_psychiatry_in_the_Soviet_Union.

2 The numbers given in the figure are observed heritability. But for the same genetic background, there can be different heritabilities in different environments. Most obviously you will not develop full-blown PKU if you are brought up in a low phenylalanine culture (e.g., low starch vegans).

3 Valproic acid has also been used to generate rat models of autism.[R34]

4 Amino acids are the primary building blocks for proteins.

5 This text is based on https://www.ncbi.nlm.nih.gov/books/NBK519704/table/ch3.t10/, which is part of the *DSM-5* text revision.[R2]

6 In rats, CORT is a slightly different compound: corticosterone. But the effects are the same and so calling both CORT makes life simpler.

7 See Chapter 5 *Emotion and internal action.*

8 Neuroticism, Agreeableness, Conscientiousness, Extraversion, Openness/Intellect.

9 ICD-11 and the DSM-5 have broken this list up into several classes.

10 The relevant *DSM-5*[R1] table can be viewed at https://www.ncbi.nlm.nih.gov/books/NBK519704/table/ch3.t13/#:~:text=anxiety%20or%20distress.-,1.,worries%20about%20real%2Dlife%20problems.

11 See Chapter 5.

12 See Chapter 5, note 16.

13 See *Thought* in Chapter 2.

14 See *Effects of mutations and neurotoxins.*

15 See *The stress response and its control.*

16 See also https://en.wikipedia.org/wiki/Attention_deficit_hyperactivity_disorder_controversies.

17 *Effects of mutations and neurotoxins.*

References

R1. American Psychiatric Association, *Diagnostic and statistical manual of mental disorders* (5th ed.). 2013, Arlington: American Psychiatric Association.

R2. American Psychiatric Association, *Diagnostic and statistical manual of mental disorders 5th ed. Text Revision.* 2022, Washington: American Psychiatric Association.

R3. Bykov, KV, Zrazhevskaya, IA, Topka, EO, Peshkin, VN, Dobrovolsky, AP, Isaev, RN, & Orlov, AM, Prevalence of burnout among psychiatrists: A systematic review and meta-analysis. *Journal of Affective Disorders*, 2022. **308**: 47–64.
R4. Cahill, L, His brain, her brain. *Scientific American*, 2005. **May**: 22–9.
R5. Carver, CS, Johnson, SL, & Joormann, J, Serotonergic function, two-mode models of self-regulation, and vulnerability to depression: What depression has in common with impulsive aggression. *Psychological Bulletin*, 2008. **134**: 912–43.
R6. Corcoran, C, Mujica-Parodi, L, Yale, S, Leitman, D, & Malaspina, D, Could stress cause psychosis in individuals vulnerable to schizophrenia? *CNS Spectrums*, 2002. **7**: 33–8, 41–2.
R7. DeYoung, CG, Higher-order factors of the Big Five in a multi-informant sample. *Journal of Personality and Social Psychology*, 2006. **91**: 1138–51.
R8. DeYoung, CG, Cybernetic Big Five theory. *Journal of Research in Personality*, 2015. **56**: 33–58.
R9. Edwards, S & Uhlenhuth, EH, Panic disorder and agoraphobia: A sufferer's perspective. *Journal of Affective Disorders*, 1998. **50**: 65–74.
R10. Gomes, FV & Grace, AA, Beyond dopamine receptor antagonism: New targets for schizophrenia treatment and prevention. *International Journal of Molecular Sciences*, 2021. **22**: 4467.
R11. Gray, JA, *The neuropsychology of anxiety: An enquiry in to the functions of the septo-hippocampal system*. 1982, Oxford: Oxford University Press.
R12. Graybiel, AM & Rauch, SL, Toward a neurobiology of Obsessive-Compulsive Disorder. *Neuron*, 2000. **28**: 343–7.
R13. Griffith, JW, Zinbarg, RE, Craske, MG, Mineka, S, Rose, RD, Waters, AM, & Sutton, JM, Neuroticism as a common dimension in the internalizing disorders. *Psychological Medicine*, 2010. **40**: 1125–36.
R14. Hilgetag, CC & Barbas, H, Sculpting the brain. *Scientific American*, 2009. **300**: 66–71.
R15. Hofer, MA, Wolff, CT, Friedman, SB, & Mason, JW, A psychoendocrine study of bereavement. Part I. 17-hydroxycorticosteroid excretion rates of parents following death of their children from leukemia. *Psychosomatic Medicine*, 1972. **34**: 481–91.
R16. Hofer, MA, Wolff, CT, Friedman, SB, & Mason, JW, A psychoendocrine study of bereavement: Part II. Observations on the process of mourning in relation to adrenocortical function. *Psychosomatic Medicine*, 1972. **34**: 492–504.
R17. Holt, P, Panic disorder: some historical trends, in *Anxiety*, N McNaughton and G Andrews, Editors. 1990, Dunedin: University of Otago Press. p. 54–65.
R18. Holt, PE & Andrews, G, Provocation of panic: Three elements of the panic reaction in four anxiety disorders. *Behaviour Research and Therapy*, 1989. **27**: 253–61.
R19. Kaufman, IC & Rosenblum, LA, Effects of separation from mother on the emotional behaviour of infant monkeys. *New York Academy of Science*, 1969. **159**: 681–95.
R20. Larsson, H, Anckarsater, H, Rastam, M, Chang, Z, & Lichtenstein, P, Childhood attention-deficit hyperactivity disorder as an extreme of a continuous trait: A quantitative genetic study of 8,500 twin pairs. *Journal of Child Psychology and Psychiatry*, 2012. **53**: 73–80.

R21. Lathe, R, Singadia, S, Jordan, C, & Riedel, G, The interoceptive hippocampus: Mouse brain endocrine receptor expression highlights a dentate gyrus (DG)–cornu ammonis (CA) challenge–sufficiency axis. *PLoS One*, 2020. **15**: e0227575.
R22. Lenzenweger, MF, Schizotypy, schizotypic psychopathology and schizophrenia. *World Psychiatry*, 2018. **17**: 25–6.
R23. Mann, FD, Waszczuk, MA, Clouston, SAP, Feltman, S, Ruggero, CJ, Marx, BP, Schwartz, JE, Bromet, EJ, Luft, BJ, & Kotov, R, A 20-year longitudinal cohort study of post-traumatic stress disorder in World Trade Center responders. *Nature Mental Health*, 2025. **3**: 789–802.
R24. McCrae, RR & Costa, PTJ, The five-factor theory of personality, in *Handbook of personality: Theory and research*, OP John, RW Robins, et al., Editors. 2008, New York: Guilford Press. p. 159–81.
R25. McFarlane, AC, The aetiology of post-traumatic morbidity: Predisposing, precipitating and perpetuating factors. *British Journal of Psychiatry*, 1989. **154**: 221–8.
R26. McNaughton, N & Corr, PJ, Mechanisms of comorbidity, continuity, and discontinuity in anxiety-related disorders. *Development and Psychopathology*, 2016. **28**: 1053–69.
R27. McNaughton, N & Glue, P, Ketamine and neuroticism: A double-hit hypothesis of internalizing disorders. *Personality Neuroscience*, 2020. **3**: e2.
R28. McNaughton, N & Gray, JA, *The neuropsychology of anxiety: An enquiry into the functions of the septo-hippocampal system* (3rd ed.). 2024, Oxford: Oxford University. https://global.oup.com/academic/product/the-neuropsychology-of-anxiety-9780198843313?cc=es&lang=en&.
R29. Noyes, R, Jr, Burrows, GD, Reich, JH, Judd, FK, Garvey, MJ, Norman, TR, Cook, BL, & Marriott, P, Diazepam versus alprazolam for the treatment of panic disorder. *Journal of Clinical Psychiatry*, 1996. **57**: 349–55.
R30. Pariante, CM, Depression, stress and the adrenal axis. Neuro-endocrinology Briefings, 2003. https://www.neuroendo.org.uk/doc/briefing_19.pdf
R31. Poulton, R, Caspi, A, & Moffitt, TE, Fear and anxiety: Lessons learned from the Dunedin longitudinal study. *Neuroscience and Biobehavioral Reviews*, 2023. **148**: 105118.
R32. Rapoport, JL, The biology of obsessions and compulsions. *Scientific American*, 1989. **260**: 63–9.
R33. Rapoport, JL, Giedd, JN, & Gogtay, N, Neurodevelopmental model of schizophrenia: Update 2012. *Molecular Psychiatry*, 2012. **17**: 1228–38.
R34. Roullet, FI, Lai, JK, & Foster, JA, In utero exposure to valproic acid and autism—A current review of clinical and animal studies. *Neurotoxicology and Teratology*, 2013. **36**: 47–56.
R35. Sapolsky, RM, *Why zebras don't get ulcers* (3rd ed.). 2004, New York: Henry Holt and Company.
R36. Stevenson, M & McNaughton, N, A comparison of phenylketonuria with attention deficit hyperactivity disorder: Do markedly different aetiologies deliver common phenotypes? *Brain Research Bulletin*, 2013. **99**: 63–83.
R37. Tandon, K & McGuffin, P, The genetic basis for psychiatric illness in man. *European Journal of Neuroscience*, 2002. **16**: 403–7.
R38. van Marwijk, H, Allick, G, Wegman, F, Bax, A, & Riphagen, II, Alprazolam for depression. *Cochrane Database of Systematic Reviews*, 2012. Art. No.: CD007139.

R39. Wang, Z, He, D, Yang, L, Wang, P, Xiao, J, Zou, Z, Min, W, He, Y, Yuan, C, Zhu, H, & Robinson, OJ, Similarities and differences between post-traumatic stress disorder and major depressive disorder: Evidence from task-evoked functional magnetic resonance imaging meta-analysis. *Journal of Affective Disorders*, 2024. **361**: 712–9.

R40. World Health Organization, *The ICD-10 classification of mental and behavioural disorders*. 1992, Geneva: World Health Organization.

11 Epilogue

Biopsychology and Your Mind explains the most basic principles underlying your thoughts, feelings, emotions, and their disorders. These mental mechanics explain how your brain's activity gives rise to your mind. It sees your emotions as key. But your emotions depend on your perceptions, thoughts, feelings, and personality, so the book covers these also. These mental mechanisms reflect activity in your brain.[1] Change your brain's activity and you alter your mind.

Mental mechanics encompasses basic principles. Some facts are required to understand them but *Biopsychology and Your Mind* limits these as far as possible. It provides a *general* foundation for understanding your mind, those of others, and the mental problems that can afflict anyone. If you understand the general principles, you have a solid foundation for further understanding of specifics.

With 86 billion neurons,[R1] the brain can seem as difficult to understand as the mind. But mental mechanics is about simple processing units and fundamentally simple systems.[2] These can be identified and studied individually. They follow rules constrained by their evolution. Critically, we can be sure of their unchanging physical nature, where the concepts of pure psychology or philosophy are much more slippery.

Understanding a brain system as a whole is helped by its stacking of neurons in parallel. You can describe, for example, the initial stages of visual processing in terms of just a few receptors in the retina and neurons strung out along a sketch pathway from the retina to the thalamus. Then, for the system as a whole say "repeat this wiring in parallel 125 million times". The size of this number is due to your eye being like a 125-megapixel digital camera – no processing steps need to be added to the original sketch pathway.

Understanding neurons is helped by their being much like other cells. Cells generally interact using chemicals as signals and electrical charges to transfer information.[3] Neurons are special mainly in terms of their wire-like extensions (axons), allowing fast communication with other neurons. An axon can be a metre long (from the spine to the big toe, for example).

But neurons do not just pass on information. Their cell bodies are information processors. They integrate information coming in and produce

DOI: 10.4324/9781003465027-11

conditional output – but in different ways than the processing units in your computer. Neurons combine simple integration (in the cell body) with simple memory storage (via changes in synaptic connections). The computational power of your mind comes from the huge number of neurons in large-scale networks.[4]

The architecture of neural systems also follows simple principles.[5] The brain does not get assembled from pre-formed parts like a computer. This is because, during evolution or development, the brain has to work properly at all stages. This constrains how it expands – with much repetition of the same basic types of change from one step to the next.

One key aspect of brain organization is that nuclei expand and then split into separate lower and higher versions – each with the same basic function and much the same processes as the smaller version. The result, when you are an adult human, is a hierarchically organized system with lower levels much like those of a simple reptilian brain or human foetus and layers added progressively above each other like marzipan and then icing on a cake.[6] This means that each layer can be studied separately before we ask how they interact with each other.

A second key aspect of your mind's architecture is that parallel, separate systems deal with each of your senses. Surprisingly, as they ascend to higher neural levels, these perceptual systems also split into separate parallel 'what' and 'where' streams that eventually meet up with an emotional 'why stream'.[7] Given that we experience an apparent mental unity, this separation of mental processes seems difficult to believe. But it explains the surprising results of brain damage, such as *The Man Who Mistook His Wife for a Hat*.[R2]

A third key aspect of your mental architecture is the functional nesting of the hierarchical organization of these parallel systems. At the periphery, sensory and motor systems are linked and produce evolutionarily old 'quick and dirty' responses – such as the withdrawal reflex and stretch reflex.[8] At each level of the ascending hierarchies, responses become slower and the processes more sophisticated. This adds higher control on top of the 'quick and dirty' mechanisms. This layering of systems also explains phenomena such as blindsight. With either brain damage or the presentation of illusions,[9] you have evidence that lower unconscious systems process information in parallel to higher conscious ones – and produce computationally different results. This makes evolutionary sense – the different results are optimal for different sorts of decisions.

The interweaving of the multiple strands of these vertical and horizontal aspects of organization is like a tartan (US: plaid) that encompasses emotion, learning, and memory and, taken as a whole, reflects your identity (Figure 11.1). *Biopsychology and Your Mind* presents an overall pattern of this tartan. At this scale, you can recognize the pattern as a person (in the way you can identify me as a MacNaughton if I am wearing my kilt). At this level, emotions are best defined in terms of the adaptive pressures that shaped their evolution[10], with interweaving of basic emotions giving rise to blends[11]

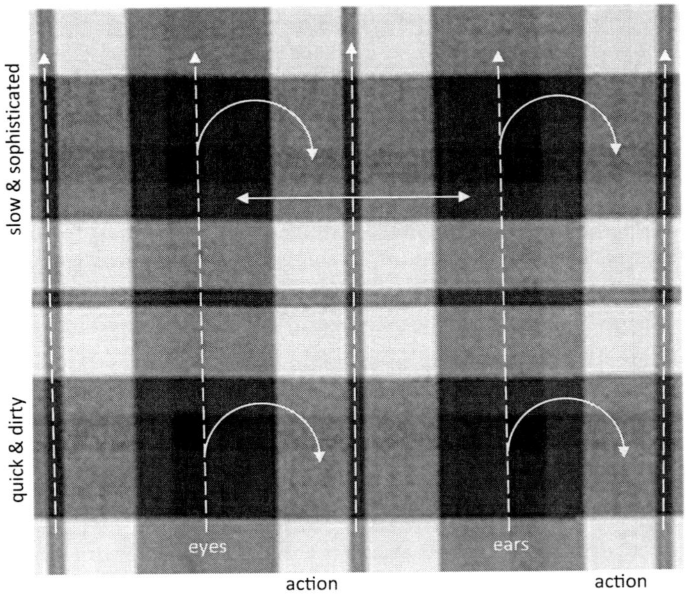

Figure 11.1 The mind as a tartan. This expands on Sherrington's "Enchanted Loom" analogy for the brain. Vertical connections are the warp and horizontal the weft. To make tartan (USA: plaid), colours are intermingled according to a fixed geometric pattern. Here (because of publisher colour restrictions) I show a grey-scale version of the simplest MacNaughton tartan – MacNaughton 'Dress' Modern (House of Edgar) – my own kilt is much more colourful – Ancient MacNaughton. If you know the patterns, you can identify a Scott in a surname kilt from the other side of a room. Your mental mechanics employ warp for separate (e.g., eyes, ears) hierarchical ascending systems and weft for integration across them. Critically, at each level, input can generate output unless a higher level takes control first. As with the tartan, the overall pattern identifies **me**. But we can go down to lower and lower pattern levels for our explanations. The book (mostly) stops before the thread level.

(as in a tartan). Learning[12] and memory[13] then solve evolutionarily novel problems (like adding a sporran since the kilt does not have pockets). Closer in (at the block and line level), the book looks at mental mechanical principles (illustrated with only a few thread-level facts). These principles allow us, for example, to see a variety of different sorts of learning as depending on just three fundamental types of neural plasticity.[14]

Biopsychology and Your Mind presents a global picture of your mind's architecture. It covers your mental mechanics in terms of fundamental principles: how it all works. But, in focusing on mental cogs and gears and wheels, it simplifies and blurs many details of the normal operation of the machine. It includes selected facts because it is intended only as a door for you to enter the vast edifice that is biopsychology. Your interests may be general or specific and shallow or deep. With more detail, I could not see how a single book could cater to everyone. So, from here, I expect you to take your own personal path driven by your own specific interests.

When you enter this building, your personal preferences may take you to just one of its many floors: neuroscience; neuropsychology; neural and behavioural pharmacology; biological psychiatry; evolutionary psychology; cognitive neuroscience; or perhaps even a side-wing like neuroanatomy. But ultimately, all these areas are interconnected; and you may simply want to wander through the whole building. Whether going deep and specific or shallow and general, mental mechanical principles should provide you with a guide to integrating the enormous amount of data with which you will be faced.

Whatever direction you go, I hope you enjoy the journey. I also hope that this book will provide you with a suitable road map to locate things on and find your way when (as often happens with brain and mind) you get a bit lost. I believe mental mechanics should help you understand your mind both when it is solving problems for you or creating them.

Notes

1 Chapter 2.
2 Chapter 3.
3 Chapter 3.
4 Chapter 9.
5 Chapter 4.
6 Chapter 4.
7 Chapter 9.
8 Chapter 9 *The perception-action cycle – spinal mechanisms.*
9 *Experiment 2.3: Parallel visual systems.*
10 Table 5.1.
11 Figure 5.6.
12 Chapter 6.
13 Chapter 7.
14 Figure 6.7.

References

R1. Herculano-Houzel, S, The human brain in numbers: A linearly scaled-up primate brain. *Frontiers in Human Neuroscience*, 2009. 3: A031.

R2. Sacks, O, *The man who mistook his wife for a hat, and other clinical tales*. 1985, London: Picador.

Index

Note: **Bold** page numbers refer to tables; *italic* page numbers refer to figures and page numbers followed by "n" denote endnotes.

For Product Safety Concerns and Information please contact our EU representative GPSR@taylorandfrancis.com Taylor & Francis Verlag GmbH, Kaufingerstraße 24, 80331 München, Germany